The Chronoscopic Society

Steve Jones
General Editor

Vol. 17

PETER LANG
New York • Washington, D.C./Baltimore • Bern
Frankfurt am Main • Berlin • Brussels • Vienna • Oxford

Robert Hassan

The Chronoscopic Society

Globalization, Time and Knowledge in the Network Economy

PETER LANG
New York • Washington, D.C./Baltimore • Bern
Frankfurt am Main • Berlin • Brussels • Vienna • Oxford

Library of Congress Cataloging-in-Publication Data

Hassan, Robert.
 The chronoscopic society: globalization, time
and knowledge in the network economy / Robert Hassan.
 p. cm. — (Digital formations; vol. 17)
 Includes bibliographical references and index.
 1. Time—Sociological aspects. 2. Information society. I. Title.
  HM656.H37   304.2'3—dc21   2003009235
   ISBN 0-8204-6707-3
   ISSN 1526-3169

Bibliographic information published by **Die Deutsche Bibliothek**.
Die Deutsche Bibliothek lists this publication in the "Deutsche
Nationalbibliografie"; detailed bibliographic data is available
on the Internet at http://dnb.ddb.de/.

Cover photo by Theodor Hassan
Cover design by Lisa Barfield

© 2003 Peter Lang Publishing, Inc., New York
275 Seventh Avenue, 28th Floor, New York, NY 10001
www.peterlangusa.com

All rights reserved.
Reprint or reproduction, even partially, in all forms such as microfilm,
xerography, microfiche, microcard, and offset strictly prohibited.

To the memory of Joe Connolly
(1938–2001),
Motherwell steelworker, 1980s flying picket,
and the person who taught me to think
through thinking about football.

Table of Contents

Acknowledgments ix
Introduction 1

Part One: Times

1. What Is Time? 11
2. Temporal Rhythms 21
3. The Commodification of Time 27
4. Globalization and the Information Technology Revolution 37
5. Digital People in a Digital Ecology 57

Part Two: Learning and Earning in the Information Ecology

6. The University in Western Society 69
7. The New Universities and the Student-Worker of the Twenty-First Century 91
8. Temporal Sense Making in the Information Ecology 109
9. The New Knowledge Production: Instrumental Versus Critical Thought 125

Part Three: The Chronoscopic Imagination: Critical Thought and Civil Society in the Twenty-First Century

10. Abbreviated Thinking 133
11. "I've No Time for Politics!" 143
12. Disorganized Capitalism and the Emergence of Chronoscopic Politics 155
13. *Amor Fati* (Embrace Your Fate) 169

Bibliography 177
Index 183

Acknowledgments

First and foremost I want to thank David Hayward, Director of the Institute for Social Research (ISR) at Swinburne University of Technology, for making this happen. He sees long-term in a world of short-termism and therefore was able to give me the time and space to research and write this. For their very helpful comments and suggestions I want also to thank Graham Ramsay in Glasgow, Tim May in Salford, Stuart Allan in Bristol and Ken Wark from Sydney (although I think he may have been in Zagreb when I emailed him the manuscript, and he then suggested alternative ways to think about what I had written). Overall responsibility for the final text, with its flashes of devastating insight and glaring errors of omission, remains, of course, with myself.

Special thanks go to Steve Jones in Chicago for his ideas regarding the original manuscript and for his faith in its ultimate publishability. Thanks also to Bernadette Shade and Christopher Myers at Peter Lang in New York for acting upon Steve's belief and putting the finished manuscript into a big abstract machine that prints and distributes books. Apart from Graham and David, I've never met any of these people, the whole process being conducted over the Internet and through email—and I'm still not sure if that represents progress or not. I also need to thank David Hudson of the ISR for his editing skills and Tania Herbert for the gritty task of formatting the text.

Very special thanks to my wife Kate Daw, artist, teacher and mother to our son Theodor. If not for them, you wouldn't be reading this. A welcome to the world, too, to our daughter Camille, born in early 2003.

Introduction

Anxiety over the perceived acceleration of life, of the speeding up of every realm of our existence, of there never being enough hours in the day to do all that we want and need to do, is not as contemporaneous with our age as one might think. In the West it goes back at least over two thousand years to the time of Plato, who fretted that a newfangled technology called writing was playing havoc with deep thinking and memory. Writing, he imagined, was akin to cheating, a kind of shorthand for thinking, a process that has its own innate temporal rhythms and one that could only be corrupted through such artificial techniques. Possibly this was the first instance of the coming together of technology with temporality, with the former beginning to influence and meter the latter—making people apprehensive in the process.

If Plato only worried about the supposedly deleterious effects of increasingly quick ways of doing things through technological means, then people, it would appear, actually got sick from it in the Victorian era. Neurologists of the time had a name for it: neurasthenia, or "nervous exhaustion," a malady resulting from trying to cope with attempting to do too many things in the unprecedentedly fast pace of life that the blossoming Industrial Revolution and modernity were generating. More recently, in the 1960s, media theorist Marshall McLuhan argued that through "electric speed all forms are pushed to the limits of their potential." This no doubt jangled the nerves of those who thought life was fast enough already or those who simply wanted the quiet life and a slower-paced existence. Even more alarmingly for such worry worts, McLuhan argued in *The Medium Is the Massage* (1967/2001:63) that "electronic technology" was rapidly shrinking the planet to the point where "time has ceased and space has vanished." Forty years ago, then, when the few computers in existence were the size of double garages and long before videoconferencing technology staged its first real-time global convention, McLuhan's influential thesis was already arguing that we lived in an age of temporal immediacy. The world, he insisted, was already a "global village."

More recently there has appeared a fresh crop of books on the subject. One of these is James Gleick's *Faster: The Acceleration of Just About Everything* (1999). In "timely" fashion, the book hooks into the fears, phobias and anxieties of life in the Information Age. The effects of computers, the Internet, flexible working, increasing interconnectivity, the obsession with "making" and "saving" time have everyone in the rich countries on a permanent "adrenalin rush," he argues. The angst (as well as the excitement), moreover, is not misplaced: the stakes here can be high. For example, on the real-time stock market, fortunes can be won (or lost) in a second; jobs can be got (or lost) in the time it takes to send or receive an email; and friends can be made (or lost) at a similar baud-rate. There is a massive paradox at the center of all this: in the "network society," we need hardly leave that beige swivel chair in front of the computer at home or at work, yet our lives have never seemed so "accelerated." People read less, sleep less and have less time for friends, for family, for relationships, for their dog or goldfish or for themselves. Gleick's book was pitched at a popular readership and sold well because it struck a chord with people, and they could recognize depressing aspects of their own lives in its pages.

In 2001, Thomas Hylland Eriksen published a similar book called *The Tyranny of the Moment: Fast and Slow Time in the Information Age*. Like Gleick's book, Eriksen's looks at the immense changes that the Information Age has brought to our lives over the last twenty-five years. His particular thesis is somewhat more apocalyptic than Gleick's wittier approach. Eriksen argues that "fast time" now dominates "slow time" in our lives and that the main culprit is our use (or abuse) of information and communication technologies (ICTs). He maintains that ICTs have created and sustained this "hurried era," and there exists the real danger of Western societies disintegrating into disjointed and incoherent fragments.

I mention these books in particular because I believe their authors are onto something, and I share many of their concerns and fascinations. They speak to something much more real than McLuhan did. For all his prescience, McLuhan theorized about a world very different from ours. By comparison, the 1950s, 1960s and 1970s were very "slow" places indeed. Our heroes and icons during these languid decades had the tempo and the worldview of a John Wayne or an Alan Ladd, whereas their equivalents today might be a particularly dissonant, and vaguely creepy, amalgam of *Itchy and Scratchy* and Bill Gates.

In terms of their technological sophistication, those earlier decades were also on quantitatively and qualitatively different planets. For example, in 1956, before McLuhan's book appeared, the first transatlantic telephone cable was laid. This three thousand mile long stretch of copper wire and rubber insulation could carry the then doubtlessly impressive grand total of thirty-six simultaneous conversations, just enough bandwidth for a couple of heads of state and a few CEOs to chat one-on-one in real-time and to hell with the cost. Global village? MTV and CNN did not exist. Rupert Murdoch was master only of the dinky little *Adelaide News* in Australia, and Bill Gates hadn't yet been weaned. Business networks existed on Rolodexes, and the Internet was only a fantastic scheme in the head of someone like Vannevar Bush or Vinton Cerf. As late as the 1970s, transatlantic telephone calls continued to be prohibitively expensive for most people. I still remember, as a youngster, nervously dialing a long-distance or overseas call, with half my mind calculating how much this was costing and the other half trying to engage in conversation. A rather less stressful, but still fairly exotic, experience during the same decade was being able to watch the 1974 and 1978 soccer World Cup finals, from West Germany and Argentina, respectively, beamed live across the world by satellite. The satellite links would regularly go down, and a red-faced local anchorman would have to wing it with some conversation on the state of play until the link was re-established. The thing is, we expected such things to happen—we were riding the technological crest of the wave, and occasionally getting dumped was part of the process. Global village? Not yet.

A couple of things have happened since Mario Kempes scored in extra time to make it 2–1 against an uncommonly lackluster Holland and put Argentina on the road to victory in the 1978 final. One of these was economic globalization, the rapid shift toward a world of open (or opening) markets, of mobile capital, of global competition and of lean and mean transnational corporations who roam the world, virtually unhindered, for the cheapest sources of raw materials, of low-priced labor and of the most lucrative markets. The other is the ICT revolution. It's fair to say the globalization could not have happened without the ICT revolution and vice-versa. These interrelated dynamics, which I shall discuss at length in this book, have formed an extremely powerful nexus that has propelled the rich countries of this world, and their economies and societies (a majority of them, anyway), into a different economic and technological plane than existed in the not so distant 1970s, a time that lives with us only in echoes through retro music and fashions.

I said also "qualitatively different" when speaking of the technological difference between then and now, and the qualitative difference is the big difference. Yes, we have the Internet, and it is capable of doing all manner of weird and wonderful things, from ordering lunch and getting a medical opinion on that rash on your child's backside, to buying a car, organizing your funeral or getting that obligatory MBA from Harvard Business School or the Online University of the Galapagos Archipelago. But the Internet is much, much more. It constitutes a digital "backbone" that connects much of the world inside a single uniform protocol within a rapidly expanding bandwidth. Hundreds of millions of people are linked in and through networks of businesses, networks of communities, networks of shared interest, and random and amorphous networks of untold numbers of individuals from every corner of the globe who are chatting, soliciting, pestering, asking, informing and, sometimes, even, scheduling to meet face to face (FTF). An example: one small corner of the Internet is taken up with MP3 file sharing, the swapping of music files through simple software programs such as Napster, Kazaa, Morpheus and a host of others; tens of *billion* unlicensed MP3 audio files have been swapped since 2001. These sites can also count how many are accessing them simultaneously, and in every visit I made (for research purposes only), there were never less than half a million people in each of them, swapping music with each other and chatting with each other. Cyberspace teems with people who intercommunicate every nanosecond of every day.

But the Internet is more than this again, potentially, *infinitely* more. The "backbone" of interconnectivity is, by its nature and design, an amorphous "thing" of almost limitless connectability. The Internet is becoming a network of infinite density and complexity. Networks of networks are building daily as new connectible devices and new users become part the digital logic that is reshaping and reorganizing our lives and our societies. In the opening passages of *The Internet Galaxy* (2001) Manuel Castells writes that:

> The Internet is the fabric of our lives. If IT is the present-day equivalent of electricity in the industrial era, in our age the Internet could be likened to both the electrical grid and the electric engine because of its ability to distribute the power of information throughout the entire realm of human activity.

This digital logic and its ever-growing suite of accoutrements, from cellphones and PDAs to email and multi-channel interactive TV, has for at least a decade been reaching into every nook and cranny of social, cultural and economic life. It is a central contention of this book that growing interconnectivity, the burgeoning of "networks of networks," is creating an unprecedented phenomenon. It is creating a technological and socio-economic nexus that has its own ecological and temporal dimensions. In other words, this continually populated, rapidly expanding and increasingly complex twenty-four hour network generates its own *ecology*, its own specific and unique environment. This is an environment that is like the natural and built environments that surround us. And, like our natural and built ecologies, we shape them and are in turn shaped by them in an ongoing, dialectical relationship. Moreover, this "information ecology," as I have termed it, generates and sustains its own *temporality*, a real-time constant present. This is rapidly displacing the chronologic temporality of clock time with a digitally compressed real-time, or what I have termed (after Paul Virilio) *chronoscopic time*.

This technologically driven shift from a society that is metered and dominated by clock time to one that increasingly is lived in the real-time present is the core of the book, and echoes much that Gleick and Eriksen have described. However, the treatment given to the subject of the "acceleration of just about everything" by Gleick and the "tyranny of the moment" that Eriksen offers, fall victim to that which they describe. For books that concern themselves with time and the effects of ICTs upon our relationship with time, they seem not to have had the time to devote to an analysis of the subject in any meaningful and illuminative way. These are "fast" descriptions of the accelerated life: non-linear, non-historical, cut—and-paste renderings that may be opened, essentially, at any page with the reader able to recognize this or that particular "snapshot" of the accelerated life.

These books are written and organized in such a way as to make the process of reading almost like surfing the Internet, a constant moving from link to link with no underlying narrative upon which to focus. What these authors wittingly or unwittingly convey is that so fast has life become, so tyrannical is the digitized moment, that we have no time any longer for deep and considered analysis, or for a reflexive and critical approach to the subject, or to be able to understand *why* we never seem to have enough time to do anything properly, or *why* we feel constantly harassed through lack of time. They give us information that we can relate to, certainly, information that we balefully recognize in our own lives, but this does not transfer into critique, knowledge and understanding. Another way of looking at these processes, and this is partly what the book will explore, is that we are *adapting to* our environments, as opposed to helping shape them in a more dialectical process, and in an environment of real-time, we are getting used to dealing in "data" and "information" instead of "knowledge." We thus become accustomed to thinking in an "abbreviated" fashion that is synchronous with our real-time processes of making sense of things through the interpretive prism of the information ecology.

In the following chapters, I seek to go beyond description and surface impressions. This book is therefore an analysis and critique of the accelerated life that has emerged though the processes of globalization and the ICT revolution. However, it consciously works against some of the effects of speed. I have structured this book in such a way as to hold its meaning and its context throughout. In that sense it is a book that goes against the logic of that which it describes and analyses, instead of being a reflection of it as Gleick and Eriksen's books tend to be. Instead of hyperlinks, screeds of more or less connected ideas that the reader can pick and choose at random, you get a linear narrative with a beginning, a middle and an end. This means taking a historical perspective, using history to help us contextualize and understand the present as well identify trends and patterns that may point toward possible futures.

The book is divided into three parts, each of which has a number of sub-chapters. Part One, "Times," is a discussion of the nature of temporality within human society. In particular it looks at the evolution and ultimate domination of clock time as a way of measuring and comprehending time in society. Indeed, so deeply entrenched is the meter of clock time within society that we need to remind ourselves that this is a social construction, not a natural, universal process. This first part of the book seeks to show how technology, time and then the economy became interlinking and interdependent processes, culminating with the final entrenchment of clock time in Western society during the Industrial Revolution. "Globalization and the Information Technology Revolution" is a key chapter in Part One. It shows how the changing imperatives of the capitalist economy inaugurated both neoliberal globalization and the ICT revolution, and how these dynamics have enabled the construction of the "information ecology." The chapter tracks the progression of the ICT revolution and globalization and how real-time has begun to displace the two-hundred-year domination of clock time in Western industrial societies.

Part Two is called "Learning and Earning in the Information Ecology." The focus here moves from analysis of the causes of the "information ecology" and the "shift from chronologic to chronoscopic time" to their effects upon the production and dissemination of knowledge in society. In particular it looks at the university and its students, an institution and group of people that have undergone what Dan Schiller (1999:144) calls a massive "makeover" due to the direct political and technological effects of neoliberal globalization and the ICT revolution. Why this particular focus? Well, the university is a traditionally central institution for the production and dissemination of knowledge in society. More, as Gerard Delanty argues, the university "is a key institution of modernity...the site where knowledge, culture and society interconnect" (2001:vii). And more again, it is also a central node in the network society, so heavily dependent have they become upon ICTs for their primary teaching, researching and administrative functions. In a process that has been termed "academic capitalism," I show how the Anglo-American universities in particular have become pseudo-businesses. These are institutions not particularly skilled in business but shiny-eyed with the ideology of the market and the supposed benefits of ICTs. As we shall see, much of this was foisted onto the universities anyway, and they must increasingly rely upon non-government sources of funding for teaching and research, orienting them away from a "universal" concern for knowledge to one focused on the narrow needs of the economy.

The students who go through these new high-tech institutions, what Marginson and Considine (2000) term the "Enterprise University," are armed with the ideas, skills, values and knowledge they have internalized there and will go on to become the influential shapers of the world in business, the media and the culture industries. This "new" student is a new species—one that has evolved to meet the economistic and goal-oriented needs of the digital times. Part-worker and part-student, they are pushed and pulled by an array of socio-economic forces that affect both what they study and how they study. Research has shown more and more students in the Enterprise University to be "disengaged" from university life, concerned in an instrumental fashion only with getting the qualification needed to put them in the running for a well-paid job. So focused are universities and students on attracting industry funding and tailoring themselves to meet industry needs that the ideas that used to be part and parcel of a "liberal education" and a "life of the mind" are downgraded and now struggle to survive.

Part Three attempts to bring all these ideas to a focus on politics and the dynamics that drive civil society in the age of information. The era of domination by clock time evolved its own politics, a chronologic politics, you might say. This form of politics was based on the clock, on the traditions of the Enlightenment and the ethos of modernity and capitalism. This section argues that the shift from chronologic to chronoscopic time has had an important effect on the way politics is now constructed and organized, both in the established politics of the major parties and in the new social movements that have sprung up around issues of globalization. Real-time networks and politics in the information ecology are changing radically the nature of politics. The maintenance of a tradition, of a sense of history and a policy orientation in anything longer than the short term, I show, is increasingly difficult to sustain. For the established parties this means the politics of "focus groups," a smug cynicism and a seriously unfunny lack of vision, and for the new social movements, notwithstanding the large "global movements for social justice" that emerged in the 1990s, "techno-politics" does not have the necessary commitment-building dynamic that personal, face-to-face politics, the politics of an earlier era, had. Civil society in the age of the information ecology, this section argues, is more and more suspended in the present, with plans, visions and ability to change society in ways that benefit the majority of people increasingly difficult to articulate and implement.

You've got this far, so if you can, read the book as it is intended to be read—from the beginning and on through to the end (with tea breaks). In so doing, in "making time" to read it all, this will constitute a small victory for "slowness," to echo Eriksen's simple and subversive advice. But also, reading this may lead to a fuller understanding of what has been unleashed upon culture and society—not to mention the production of knowledge by the negative nexus of globalization and the ICT revolution. Finally, this shortish book (they have to be short nowadays to see the light of print) will hopefully show that the processes I describe are fundamentally *social* and that only through democratic and inclusive *social action* informed by an understanding of the world and its dynamics will positive changes emerge.

Part One

Times

Chapter One

What Is Time?

> *It is, in fact, extraordinarily difficult to think and talk about time.*
> Barbara Adam, *Timewatch*, (1995:5)

> *What, then, is time? If no one ask of me I know; if I wish to explain to him who asks, I know not.*
> St Augustine, Bishop of Hippo

"Theo, what is time?" I asked my five-year-old son one day over breakfast. I half expected a light-hearted or flippant remark, as is his style—or something similarly unusable as a way to begin a book. His immediate answer, reinforced with a jabbing, butter-coated finger, was: "Time is what's on the clock, Dad!" Theo wasn't being glib or offhand, just slightly annoyed at the apparent obviousness of the answer to my question.

Theo's response proved a point, though. On one level, the answer to the question "What is time?" *is* rather obvious; time as represented by *clock time*—the experience of *duration* through seconds ticking through minutes and hours, regulated by a mechanical or electronic device—is an abiding human relationship with one form of temporality. It goes deep and wide into the many elements of culture and society, its institutions, languages, schedules, habits and the many and varied forms of relating to one another and with the world. As Arran Gare has noted, "The permeation and domination of life by abstract [clock] time has become so complete that it is difficult to realize just how extraordinary this is" (1996:104). Consciously and subconsciously, clock time influences and shapes much of what we do on a daily and hourly basis. From the moment the radio alarm or clock bell wakes us from sleep, our profound relationship with clock time is once more renewed. It re-emerges into our life as a prodder, motivator, scheduler and organizer. From the moment the timepiece sounds, we make a series of conscious or subconscious calculations that are based, developed and honed through constant routine. Somewhere in the dark recesses of our minds we almost automatically determine how much time there is "left" before we need to be up and out of bed, get washed and dressed, eat breakfast and leave the house for work or school, appointments and so on. All through the day, clock time, with its unerring electro-mechanical persistence, serves as a powerful underpinning of our actions, thoughts and plans. From the idle luxury of thinking about how to "spend" our "free" time, that ever-diminishing time we have to ourselves, to the conscious planning or "timing" to obtain the most out of the minutes and hours of the day, time "occupies" us in a very real way. We spend our days "killing" time, "wasting" time and "making" time in our ongoing and deeply internalized relationship with this abstraction-made-real in the shape of an electro-mechanical device. Its torment (or its pleasure) ends when we fall asleep, only to re-emerge once more the next day.

Clock time regulates, categorizes, organizes and schedules not only individuals but also their interactions with other people and society in general. From the instant we are born, a clock-driven bureaucracy fixes us in time and space for all time. Our birth certificate records, to the minute, when and where we came into the world and our death certificate will indicate when and where we will have left it—with a similar attention to detail and accuracy. This will be our time-life span. There is a date, day, hour and minute in that clock time to come that will be the precise time of our death. This time-life span will become a statistic that will go to make up many other statistics, such as the average life span of an adult in a Western society. Clock time regulates, organizes and schedules the principal institutions of our modern societies—and synchronizes the individual to their tempo. Timetables inform us when to be at the bus stop, the train platform or the airport. Clock time and experience tell us when to avoid certain roads at certain times of the morning and afternoon, the so-called "rush hour," when harassed people with cars respond to their own synchronized clock time imperatives. The primary imperative is, of course, work time. Work time is far and away the most researched aspect of time. There are good reasons why research on the subject of work time is so large and research on other aspects of time such as time and health, or time and physiology, learning and memory is so scarce. In modern, industrial and/or post-industrial societies, work time is the core time around which much of our active life revolves. Work time and the rationalized forms of industrial production it engenders are the major organizing and disciplining forces in modern society. Work time gives the individual, society and the economic structures that sustain these regularization, standardization and predictability. Local, regional, national and international economies are developed and organized around the time economy of the clock. Without clock time, modern industry and the ways of life we so take for granted would be impracticable. Lewis Mumford recognized this almost seventy years ago when he wrote that the key technology of the Industrial Revolution was not the steam engine or the railway—technologies that connote the adventurous works of heroic men grappling "successfully" with nature—but the monotonous persistence of the not very exciting mechanical clock (1934/1967:14).

Daily work time commonly revolves around tasks that are themselves organized around the logic of clock time. For the individual at work, hardly any task, in any job, does not come with an implicit or explicit timeframe—a time for completion or a time for deciding that too much time has been spent on the task before abandoning it. Machines such as computers, lathes, microwave ovens and cars perform specific tasks in a specific time, because they were designed with such a time-function in mind. Indeed, modern technologies, according to Hörning et al., are "time-loaded," that is to say; they are designed and developed with the exigencies and imperatives of clock time and work time built into them (1999:293–308). Humans build clock time into technologies that, through constant use of them, synchronize their actions with the machine's "time-loaded" function in an ongoing dialectical relationship. Computers and computerization are perhaps the most illustrative contemporary examples of the clock time/work time process. Mumford's insight into the importance of the mechanical clock for the success of the Industrial Revolution has recently been overtaken by David Boulter's analysis of the significance of the computer. "It makes sense," he writes:

> to examine Plato and pottery together in order to understand the Greek world, Descartes and the mechanical clock together in order to understand Europe in the seventeenth and eighteenth centuries. In the same way, it makes sense to regard the computer as a technological paradigm for the science, the philosophy, even the art of the coming generation. (1984:109)

The so-called "information technology revolution" and the effects of computerization upon society, especially in the realms of higher education and knowledge production, are central concerns and will be discussed at length later on in Part One and constitute a primary narrative strand in Parts Two and Three. It may be useful at this point, however, to preface those discussions with some remarks on the importance of computers and their logic as they pertain to clock time, work time and the "accelerating" pace of modern life.

The information technology revolution is primarily an outgrowth of the old Industrial Revolution in that both are concerned with much the same things and much the same processes: production and consumption, innovation and automation, competition and profit, distribution and economies of scale. Much like the nineteenth-century capitalists who saw the steam engine and the railroad as the latest, most innovative and most efficient way to produce and compete in the marketplace, the contemporary capitalist is using ICTs to do the same thing. We can trace the genesis of the information technology revolution to the late 1970s, the period of the so-called "crisis of capitalism" (Harvey, 1989). Throughout the decade, intense competition within world capitalism had progressed to the point where the system as a whole was teetering at the brink of catastrophe. The production régime of "Fordism," that is to say, one based on long production lines churning out standardized items for a "mass society" to consume, had reached a point of maturity and unworkability (Harvey, 1989:125–273). The industrial, political and technological logjam caused by this maturity and immobility was eventually broken through the application of new technologies of computerization and automation and "flexible" systems of production such as the "just in time" method where overproduction and wastage are kept to a minimum.

Computers and flexible production systems rapidly spread across almost all industries, providing a boom for the computer industries themselves which, spurred by intense competition within that industry, drove the information technology revolution faster still. The over-riding logic quickly became a race to develop the *fastest* computers, thereby reducing the time-gauge of industries and societies to nanoseconds (one billionth of a second). This was and still is no revolution unambiguously in the service of the people—or even of the IT industry itself. As Neil Postman has noted, the paramount logic of the information technology revolution is no longer about developing and using computers and computerization to solve social, economic and production problems (if it ever was). The principal "problem" the revolution seeks to solve, he observes, is how to store, generate and process information at faster and faster speeds. Speed has been raised "to a metaphysical status" and is now "both the means and end of human creativity" (1993:61). Moreover, the concomitant obsession with achieving ever-greater speeds in information processing and industrial production has, according to Paul Virilio (2000:118), resulted in:

> The acceleration of real-time, the limit-acceleration of the speed of light, [that] not only dispels geophysical extension...it dispels the importance of the longue durées of the local time of regions, countries and the old, deeply territorialized nations.

A central consequence of the information technology revolution, then, has been the lightning-paced spread of computerization throughout society, with the rhythm of clock time that drove the Industrial Revolution now being accelerated by the infinitely more rapid "time-loaded" functions of computerization.

Time as a Social Construction

After reading this brief outline of the centrality of clock time to our lives today, it may be somewhat easier to understand Theo's response to my question, "What is time?" His short life and his child's world have revolved in a profound way, as with most children, around the regime of clock time. Clock time is therefore a very important influence, and children have to grasp it quickly—if only because it suffuses their life and their environment so much. Breakfast time soon becomes reading time, which slips into play time, then dinner time, moving into "quality" time with parents, on to bed time and so on. Calendar time marking birthdays looms large in the imagination of a child, too, as do the calendar's festive days such as Christmas or Easter, ancient Christian festivals that have been turned by commercialization into children's festivals, a time for celebration and consumption, with the emphasis very much on the latter. Teaching children to "tell the time" is considered by parents and schools to be as fundamental a life skill as learning words and numbers. Alongside learning how to tell the time at school, punctuality, or the merit and necessity thereof, is similarly and monotonously dinned into impressionable heads. For most of us, this skill of temporal self-organization soon becomes a deeply internalized process, preparing us for the adult life of work where punctuality and predictability in habits and actions are essential (Nowotny, 1994: 63). Clock time, then, from the earliest years of our lives has become such a deep part of our existence that we alternate from being barely aware of it to being acutely conscious of it. It is at once oppressive and comforting, a source of exhilaration or panic or something that gives us a sense of identity or feelings of alienation, and it may easily be all of these in the space of a couple of minutes.

Up until this point we have been discussing clock time as "time." But, in fact, as Barbara Adam argues, clock time "is governed by the nontemporal principle of time" (1995:52). Clock time is a *form* of representing time and is one that is radically different, as we shall see shortly, from the many other temporalities that interpenetrate our lives. Moreover, clock time and other conceptions of time are social constructions: time exists only in our minds. "Time is 'in' the universe; the universe is not 'in' time," as Paul Hewitt (1974:515) has argued. Clock time, for example, is a form of time that has been abstracted from its "natural" source and become "machine time." In fact, it is arguable that clock time isn't really "time" at all but a simple mechanical "metering" of the vast diversity and contextuality of human experience and of times in the universe.

Clock time can be viewed as the measurer of the immeasurable. Consider the statement that *calendars and clocks (chronological time) combine time as a measure as they seek to measure time*. This constitutes a double paradox that has to be considered if a fuller understanding of the complexities of time is to be appreciated. The clock is designed to the principle of invariance, that is, a regularized, precise metering, irrespective of context. Yet context(s) are characterized by a fundamental *variance*. For example, the hours of daylight change, imperceptibly, every day, yet chronological time measures each day's passing as exactly the same. Time that is calculated upon the constellations and movements of the stars, so-called "ephemeris time," is also variable, as stars do not reappear in the same astral position every calendar year. More readily perceivable, perhaps, is the fact that in the Western Gregorian calendar not every year consists of 365 days, with "leap years" occurring every four years. Indeed, clock time measuring is itself subject to variance, a variance brought about by natural forces, but we nonetheless act on the premise that it is precise and unerring. "Clock time may be designed to the principle of invariance," Adam writes, but:

> it is subject to environmental influences such as gravity, but we tend to disregard that invariability and focus on the invariable abstraction: the idea of the second, the minute and the hour. Unlike the variable rhythms of nature, the invariant, *precise measurement is a human invention* and in our society it is this

created time which has become dominant to the extent that it is related to as time *per se*, as if there were no other time. (1995:24–5)

The idea of time as represented in clock time is an abstraction that is deeply embedded in our culture and has a long history. It is based upon a mechanistic and materialistic view of the world that has its roots in the very origins of Western thought. This lineage stretches from ancient Greek philosophy to Isaac Newton and Adam Smith and was concerned primarily with the idea that nature in all its workings is comprehensible through the particularly human concepts of measurability, calculability and controllability through conscious intervention. This tradition gave birth to modern science, whose very *raison d'être* as a rational discipline is to seek to understand and control nature for human, goal-oriented purposes. Modern science, through underpinning the scientific worldview with mathematics, radically distinguishes itself from a philosophy of nature that would take as its premise a world and a reality that is complex and diverse. The science-based perspective generates a particular type of information and knowledge whose data are facts drawn by our senses or instruments from the world of nature but whose intelligibility is *mathematical intelligibility*. Consequently, the characteristic approach to reality peculiar to science considers only phenomena that can be observed and measured. The ways through which observation and measurement are to be achieved, and the more or less unified mathematical reconstruction of such data, constitute the context for meaning for both science and scientist. This unified mathematical conception of reality also has very deep roots in Western culture. Arithmetic, for example, was developed in ancient Greece alongside commerce as exchange values that were measured quantitatively in terms of coinage. Pythagoras extended arithmetic well beyond the basic needs of commercial transactions, however. As Gare (1996:76) describes it:

> [Pythagoras'] success in describing geometrical figures in numerical terms and in finding simple numerical ratios in the intervals in a string producing consonant harmonies led him to conceive of all things as number. These numbers...performed the same structural role in culture as the sacred realm in previous times.

Modern science based on the principles of mathematics, regularity, measurability and predictability paved the way for the development of increasingly sophisticated technologies, primarily the machine, and particularly during the rise of mercantile and industrial capitalism. The economic competition inherent in capitalism intensified the struggle to develop machinic "solutions" to economic and social problems still further.

In the late seventeenth century this "mechanistic" view of the world and its ordering was given major theoretical and scientific credence by the work of Sir Isaac Newton. In his *Principia*, published in 1687, Newton developed his theory of gravitation and worked out the movements of the planets and stars based on the interrelationship (as he saw it) between time and motion. Importantly he believed time was abstract and *external* to the world. In his famous definition at the beginning of his *Principia*, Newton declared "Absolute true and mathematical time, of itself, and from its own nature, flows equably without relation to anything external." He regarded moments of absolute time as moments that follow a continuous linear sequence. The rate at which these moments succeeded one another is independent of the universe and its processes (Whitrow, 1972:128–9). The most powerful legacy of Newton's work was that it gave an abstract, mathematical and mechanistic foundation to perceptions of how the natural world and its place in the universe are constituted. Indeed, in keeping with the emerging thought of Enlightenment philosophy that advocated rational science and technological development as evidence of human progress, the machine and in particular the clock became a metaphor for the world and its rational, harmonious ordering. The idea that the universe could be mechanically represented by clockwork became a powerful one, with an ancient lineage stretching back to medieval times. This was given physical form in the clockwork machines of Galileo that showed the movement of the planets around the sun, powerfully implying that the universe itself is a clocklike machine. Consequently, since the beginning of the seventeenth and eighteenth century period of Enlightenment, the world came increasingly to be conceived and comprehended as machine-like, producing a potent worldview that has shaped the Western intellectual tradition and resonates powerfully still.

Another major Enlightenment figure deeply influenced by Newton and his mechanical conceptions of the world, and whose ideas (or a particular slant on them) continue to shape economic and social theory today, is Adam Smith. His writings, especially *The Wealth of Nations* (1776), were conceived, written and published during one of the most momentous periods in world history—the beginnings of the Industrial Revolution. This was the age of tremendous technological innovation that had evolved out of the revolution in science. Smith's most abiding concern was the workings of the economic processes that led to wealth creation in society, what was termed *political economy*. Like Newton, he believed that reality, in this case the workings of the economy and society, could be scientifically measured, calculated, comprehended and predicted. The economy, he argued, worked like a machine where the "laws" of supply and demand would produce an optimal equilibrium—the famous "invisible hand" of the market that lay behind the generation of society's wealth. Just as Newton abstracted time from the reality of the universe, Smith abstracted the workings of the economy from the realities of human society. His was a deeply moral philosophy, where individuals were required to play their part *vis-à-vis* the mechanical working of the economy, just as individuals are required to adapt and synchronize their actions with the abstraction of clock time. This was the intellectual locus of classical liberalism and modern conservatism, ideologies that permeate and dominate political and economic thinking today. Describing the essence of this worldview, Gare (1996:143) observes that "A good machine is one which functions efficiently, and for the economic machine to function efficiently, individuals must accord with the image of humans presented by [Adam] Smith." The clock, and by extension the machine, had thus become the primary metaphor and abstract symbol for the Western orientation to the reality of the world.

Today, notwithstanding the advances in the sciences of thermodynamics, relativity theory and chaos theory, branches of science that indicate the enormous complexities, diversities and temporalities that exist in the universe, the power of linearized clock time still deeply inserts itself into people, into cultures and into societies. This abstracted clock time is essentially inert time that recognizes no difference in the world or the individuals and societies that populate it. It cannot discriminate between times and cannot measure changes in times. For the clock there is no difference between seasonal change or environmental change, and its metering is blind to the energetic temporalities of heat and cold or of the social-cultural times of the sacred or secular. Barbara Adam (1998:40) puts the case succinctly and somewhat caustically:

> Even today the sought-after ideal of Newtonian physics is independence from time and space, an atemporality which renders change static. The process by which this understanding is accomplished is one of extracting parts from their interactive, interconnected wholes, seeking the unit-part in its ultimate simplicity. That achieved, motion is not only measurable and predictable but also reversible. That is to say, if one removes friction, if one excludes gravity, if one excludes electro-magnetism, if one excludes interaction, if one excludes context and boundary conditions, if one excludes life, if one excludes knowledge, if one

excludes any kind of human activity, emotion, interest and frailty, if one excludes all that, then one is left with a universe of perfect symmetry, single parts in predictable, uniform motion and reversibility....

In the medieval period, people saw time as relational and dialectical, as opposed to absolute. Perceptions of times were influenced by the cosmos, by tides, by seasons and by religious observances, what Nigel Thrift has called "the rich temporal patchwork of the ecclesiastical round" (1996:180). The "progress" of the Enlightenment and of science and technology has taken this relationship away. We have undergone what Marshall McLuhan described as a "metaphorical amputation," and into this metaphorical void rushed clock time. Moreover, as Nowotny (1994:56) reminds us, notwithstanding the growing awareness and understanding of other temporalities, complexities and diversity, clock time remains today the "socially prevailing" time. Not only are we continually compelled to adapt to the mathematical metering of existence through the abstraction of algebraic, linear clock time, the abstraction itself works upon us, synchronizing individuals, industries and societies to its precision metering. So deeply has this process inserted itself within us that we *seek out* the rhythm of clock time, internalizing the very temporality that narrows our horizons and confines the many possible ways of being and seeing that other temporalities may offer. In other words, it is sometimes possible to find reassurance as well as comfort in clock time even as we repress and re-regulate other temporalities through it.

It may be useful at this point to discuss that which clock time constantly displaces, represses, colonizes and replaces, what Adam has termed "the multitude of times that interpenetrate and permeate our lives" (1995:12). Most of us are at least intuitively aware of the many other temporalities that exist both in us and in the natural world that surrounds us, such as the rhythm of the seasons or the differing physical tempos of pregnancy in humans or other living creatures. However, these are continually repressed and colonized by the straitening logic of the clock in our daily lives. To bring these times to the forefront of our consciousness will allow us to more fully comprehend what it is we are losing and, importantly, how the *computerization* of clock time continues to disengage us from the natural world of which we are part.

Chapter Two

Temporal Rhythms

It was noted earlier that the life of the child is suffused to an ever-increasing degree with the meter and constraint of clock time. Nevertheless, through simply living in the natural world inside one's own body, interpenetrative temporalities constantly impress themselves upon us in countless conscious and unconscious ways. As a child, these reminders and impressions of other times are largely incomprehensible and come to us as sense impressions which we have difficulty, usually, in understanding. The increasing dominatory effect of clock time upon society ensures that interpenetrative temporalities in our body, of consciousness and in the natural world, remain difficult to experience fully and understand more easily. This accounts, partly, for the extraordinary difficulty all of us experience when trying to think, talk or write about time—as the quotes at the beginning of Chapter One attest to. Rather, we tend to *intuit* these times, a process possibly originating from very ancient human cognitive faculties that were once a more prominent part of what Walter Ong (1967:16) called the "sensorium," that "total complex of awareness by which man earlier situated himself in his lifeworld." The sensorium could be seen as the result of the relatively unmediated dialectical interaction between ancient human beings and their environment. The human senses would organize and develop in response to environmental (and cultural) needs and pressures, with the most necessary senses (in the context of the local natural and cultural environment) being prominent in the sensorial composition. The actual arrangement of the sensorium, of course, would be subject to change as changes took place in human cultures and societies and in the natural world in which these cultures and societies existed. However, with the coming of clock time as the dominating and prevailing temporal rhythm in our social, cultural and economic spheres, our sensual awareness of other times has been repressed or dulled through the internalization of clock time.

What are some of these "other" times, other rhythms, awareness of which lie in us only in residual or dormant form? The difficulties of talking and thinking about time may be eased somewhat at this point by resorting to a short anecdote based on personal experience. My own personal experiences are of course unique, but the feelings the story describes are undoubtedly fairly common—if not universal.

Extra Time

I have vivid memories, as a child growing up in Scotland, aged around seven or eight, of watching a soccer match being played in the grounds of my local primary school. It was a cold and dark afternoon, late in January. A freezing wind with squally rain in it continually lashed the footballers on the park and continued on toward the massed spectators, hitting us all squarely in the face as it blew on up the slope. My school was matched against the local rival school, and so attendance—in our prescribed role as noisy and partisan supporters—was deemed mandatory by our schools' authorities. It was a semi-final match, and so our attendance was considered doubly important. The soccer pitch was set in the side of a hill with one side of the playing area being banked naturally by the hill itself and the other side sloping steeply away, making viewing from that side of the playing surface impossible. I remember thinking in a vague and childish way that this was an inefficient way to construct a soccer ground—a whole long section of the ground was effectively off-limits to spectators.

I began to rouse from my distracted reverie on the lack of commonsense in the layout of the soccer pitch by the increasing wind and rain—and then, gradually, by the game itself. It began to dawn on me that I was daydreaming, and the reason I was daydreaming was because the game was, I felt, exceptionally uninteresting. I had been standing watching in the wind and rain for nearly an hour and a half and by now had had enough. The game was dreary in part because of the weather and in part because I did not want to be there—and it seemed like an eternity to me. Suddenly someone close to me shouted in the general direction of the players: "There's only five more minutes left!" A few faces from both teams turned briefly toward the source of the bellowing and then continued their participation in the game with a new sense of urgency. I remember thinking that this change in them was rather startling—it was physical and immediate. Something unintelligible to me had happened, and I could only vaguely comprehend it in terms of that we (those on the park and myself) must be having different experiences of the same thing. Our perceptions of time were different. They were very different. I was cold, wet and hungry, in a place I did not want to be and watching a game that I found boring. Time passed slowly and seemingly interminably. I did not have a watch, but if I did, I would have been almost convinced that the hands were moving slowly and would have been winding up the mechanism to near breaking point. The shout "There's only five more minutes left!" seemed to me a needless cruelty—five minutes in my state of mind seemed a long time indeed.

A soccer game lasts ninety minutes, but its relationship to clock time begins and ends there as within the game there are many differing temporalities being experienced. For the twenty-two players taking part, the remaining five minutes was time that was slipping away—*rapidly*. The game was even at one goal each at this point, and so the sense of emergency and the impression of lack of time affected both teams. This new temporal dynamic, one that had been initiated by the realization that the game was soon to finish, had manifest aural and physical dimensions. The noise level from the players and the spectators increased dramatically, and the physical action on the ground visibly quickened as players on both sides redoubled their efforts to get what would probably be the winning goal. In any event, at the end of the ninety minutes, the scoreline was still even, and as the game was in a knockout competition, an extra thirty minutes was allotted, as the rules stipulate, to try to get a winning result from one team or the other. Suddenly, when the whistle blew for the end of "normal time," the temporal dynamics of the game were shattered once more and reorganized into several new dimensions.

For many of the players an extra thirty minutes was the last thing they wanted. From being acutely aware that time was running out, many players were now reckoning that there was *too much* time. The physical effort of running, tackling and kicking a ball for an extra half-hour can have a major impact upon the game's outcome. Legs and lungs begin to ache, and so running and breathing gets harder. Mistakes can easily be made through fatigue, but the thirty minutes still need to be played out—and this now seems like an eternity. Tiredness, the possibility of defeat and endless minutes stretching out into the future replaced the sense of time running out and the frantic physical action that had been the experience of just moments earlier. For the spectators, the sense impressions of temporalities were no doubt mixed. For those, like myself, who found the game tedious, an extra thirty minutes was seen as more unwanted and seemingly unending time. For those who had been enjoying the game and thought that their team had a good chance of winning, thirty more minutes constituted another course in the feast, something to be savored and enjoyed. For those spectators who thought their team might lose in the fitness stakes, extra time came with a sense of foreboding and the feeling that it all may go badly wrong.

This story of how we relate in different psychological and physical ways to the rigid meter of clock time was concerned with me as an individual. But of course the individual does not live in a vacuum; he or she is a member of a particular society and a participant in a specific culture. These have temporalities that we all share and comprise a diversity of other times that humans can experience.

I have tried to describe and explain something that most of us intuit anyway: the existence of various other temporal rhythms that permeate and interpenetrate our lives. It is impossible to understand time as independent of social processes. Physicists have for a long time argued that time did not exist before matter, so we are able to argue, too, that time did not exist prior to material processes. All forms of temporality, like clock time itself, are social constructions and, as social constructions, are subject to human intervention and change. Technologies of medicine can alter our circadian rhythms; night can be turned into day; our time-life span can be and is being lengthened through advances in medical science, and the temporalities of women's reproductive systems are subject to increasing intervention and alteration. The discovery of properties of time in the material world through physics, biology and geology are important as they afford us choice, in theory, as Harvey puts it, "as to which particular process or processes shall be used to construct time and space" (1996:211). Harvey here speculates upon an almost perfect scenario, one where no single temporality dominates material and cultural processes in the world. In such a world, we would have much more control over and autonomy within the various temporalities we construct. However, as I have been arguing throughout this chapter, industrialized clock time permeates most societies, obliterates or displaces other temporalities, and increases its metronomic grip upon our culture and societies with every new technological invention which has clock time built into its design, construction and application. Clock time, too, has undergone its own process of outside intervention. The social construction of clock time, something that we consider almost a fact of nature, a force outside human control due to its domination over us, is being reconstructed by ICTs. Karl Marx in the *Communist Manifesto* of 1848 observed that technology in the period of the first Industrial Revolution compressed space. In the so-called "third Industrial Revolution"—the ICT revolution—technology is compressing time as well as space. And the significance of the compression of such a deeply internalized process, as will be discussed in later chapters, cannot be understated.

For now, however, I shall underpin that later discussion by looking at the "commodification of time," or how clock time became inextricably linked to the processes of industrialization, of social and cultural domination and of capitalism.

Chapter Three

The Commodification of Time

A clock is a lock.
Balkan proverb

We have seen how clock time has, in an extremely powerful way, insinuated its abstract metering into our lives, our cultures and our societies. We have seen also that it has displaced and dulled our sensitivity to other temporalities that exist in the natural world and in our very bodies. Today, the power and the disciplinary and regularizing effects of clock time are indisputable. So deeply infused into our being is it that we need to stop and think about it in order to be aware of it. Except when time is "pressing," and increasingly this is a lot of the time, we are hardly conscious of its existence, so "natural" has this abstraction become. Indeed, my use of the word "time" in the second instance in this sentence is an example of our unthinking and automatic reference to it. In fact, to use a term by Jürgen Habermas (1987:336), the abstraction of clock time has become a "real abstraction," one that paradoxically has the effect of being both illusion and reality at the same time. The modern world "runs" on clock time, and without its synchronizing effects upon peoples, industries, markets, distribution chains, consumption and production patterns etc.—what E.P.Thompson called the global "work-rhythm"—the economic and social world as we know it would soon collapse into chaos.

Whilst the preceding discussion dealt at some length with the fact of the domination of clock time in our world, it did not explain *why* this is so. We have discussed the internalization of the dynamics of clock time, but what were the external forces which made internalization both possible and, arguably, irresistible? Why weren't people and cultures able to oppose this logic, to stay more in tune with the myriad irregular rhythms of life? Why do we no longer appreciate fully the approximate rhythms of the seasons and the circadian rhythms of our bodies? Why are the time(s) of the Nuer of Africa, the Cherokee of North America or the Australian Aborigine so alien to us, and why do we simply (and rather vaguely) intuit other temporalities instead of living and experiencing them more completely? To understand the nature of these external forces—the external forces that have contributed to our deeply internalized relationship with clock time—we need to go back once again to the period of the beginning of the Industrial Revolution and the start of the factory system of production. We need to begin with an appreciation of the nature of *value* and of the *commodity*.

The Value of Time

What is "value"? Again, rather like the question to my son Theo that began Chapter One of this book, "What is time?," it may seem on one level a fairly obvious and, on another, profoundly difficult question. Value, one may answer if pressed, is something that you are prepared to pay for or something that you "put into" a process or a "thing" to make it "valuable." You will give money for something only if you think it has this nebulous quality. Think of going along to a car yard, buying an old banger and working on it until it becomes a shiny, souped-up street machine. This is, in today's parlance, adding value. We can grasp this process. But what *is* value, what is it that we put into the process or the "thing"? "Our time," we might say. And this is partly right, as we shall shortly see—but why is our time valuable? Once more the question becomes difficult and somewhat circular.

As Europe was emerging out of a thousand years of feudalism and a new capitalist class was beginning to assert its power over the old aristocracy, these very questions began to exercise the minds of many of the prominent philosophers of the dawning Age of Enlightenment. As industrial capitalism and the factory system began to be the dominant mode of production in place of agriculture, these sorts of questions became more pressing. Adam Smith, whom we met earlier, was one of the first philosophers of the eighteenth century to give the question of value serious treatment. For him, value or "wealth" was created in society through the innate disposition of individuals toward self-interest, a "certain propensity in human nature...to truck, barter and exchange one thing for another" (1776/1965:397). According to Smith, and to later disciples such as Margaret Thatcher, visceral self-interest is what constitutes society, with society itself being simply agglomerations of people interacting together, each for him or herself, primarily. Smith gave to this perception of reality what many neoliberals today have seen as a blinding piece of insight into human nature: "It is not from the benevolence of the butcher, the brewer, or the baker that we expect our dinner, but from regard to their own interest" (1776/1965:324). This view of humans provided him with the basis for what he took to be an "objective" or "scientific" theory of value:

> The real price of everything, what everything really costs to the man who wants to acquire it, is the toil and the trouble of acquiring it.... What is bought with money or with goods is purchased by labor as much as what we acquire by the toil of our own body.... They contain the value of a certain quantity of labor which we exchange for what is supposed...to contain the value of an equal quantity. (1776/1965: 30)

The important idea that Smith introduced into the question of value was that of human *labor power*. It accorded with his conception of people as fundamentally self-interested individuals who would not conceive of giving something freely—except in a special form, the form of a gift. Smith's thoughts on the nature of labor power and value more or less ended here. It was not until the nineteenth century that a "labor theory of value" was developed by, primarily, David Ricardo and Karl Marx. Ricardo was very influenced by the works of Smith and viewed the question of labor power and value in similarly abstract terms. Why is labor power valuable, notwithstanding Smith's self-interested conception of humans? Ricardo in his 1817 treatise *On the Principles of Political Economy and Taxation* argued that the amount of *labor time* embodied in a particular commodity, that is, that amount of time spent working on the production of a commodity, gave the commodity its value. This was a short sidestep from Smith's theory and still constitutes a somewhat vague abstraction which only unclearly posited a connection between labor, labor time, and the production and the value of commodities. It saw labor time and labor power as something external to the production of commodities. It still did not give a concrete and socially rooted underpinning to a theory of the production of value. This came from Marx some fifty years later.

In Volume One of *Capital* (1867) Marx follows Smith and Ricardo by pointing out that there exist two forms of value: *use* value and *exchange* value. The former is value that exists in the production of items for individual, private consumption or use. For example, if one is homeless, then basic shelter has considerable use value. Once housed, the use value of other shelters one has access to is greatly reduced. They only gain value as items of *exchange* in a marketplace and thus under certain historical conditions become imbued with exchange values. For Marx, these conditions are, of course, commodity production under capitalism (Marx,1967:125–244). Where Ricardo and Marx differed is whilst Ricardo (after Smith) saw labor power and labor time as something abstract and external to the production of value, Marx saw these as absolutely central. Capitalism, commodity production and the conditions of this commodity production (the social relations that brought forth value and wealth) are deeply rooted in the social and material relations of society. And for Marx this was one of unequal exchange relations, where the owners and controllers of capital exploited waged laborers who had nothing else to sell except their labor. Labor had become the central commodity in the production of a world of commodities. The usually neglected subtitle of *Capital* is *A Critique of Political Economy*. It was this critical analysis of the market and its relations, something Smith and Ricardo saw as "natural" and not fundamentally *social*, which constituted the great breakthrough that Marx made in the analysis of value and the commodity.

Time was also a central factor in this exploitative relationship. Indeed, without the dynamics of clock time metering and measuring, the production of exchange value-imbued commodities for the marketplace would be impossible. Prior to the rise of capitalism and commodity production under the factory system, clock time had only a tangential relationship to the cultures and societies of Western Europe. The other, more irregular temporalities of life still held sway for most people: the seasons, night and day in their always-changing durations, sleeping and waking were all fairly fluid and localized dynamics. Timepieces such as clocks and pocket-watches in the seventeenth and much of the eighteenth century were of questionable accuracy and were bought and worn mainly as symbols of social status. They were elaborately decorated and encased in precious metals such as gold and silver and accessible only to the rich. The work-rhythms that dominated the lives of peasants and workers at this point in history were "task oriented" in that they corresponded with the "natural" rhythms of life, such as milking cows, tending sheep, sowing crops and harvesting fields. This task orientation, as E.P. Thompson has shown, has at least three distinct features that set them in stark contrast to the work-rhythm of the clock:

The Commodification of Time 31

> First, there is a sense in which it is more humanly comprehensible than timed labour. The peasant or labourer appears to attend upon what is an observed necessity. Second, a community in which task-orientation is common appears to show least demarcation between "work" and "life." Social intercourse and labour are intermingled—the working day lengthens or contracts according to the task.... Third, to men accustomed to labour timed by the clock, this attitude to labour appears to be wasteful and lacking in urgency. (1967: 60)

As the Industrial Revolution took hold, the task-oriented approach to work and the temporal worldview it sustained began to disappear. As we have seen, the Industrial Revolution had been underscored by the revolution in science and the rise of rationality through the triumph of Enlightenment thought. Rationality and the organization of production were central, and clock time was at the heart of these processes. As Adam Smith had shown in his famous analysis of a pin factory, tasks were performed far more productively if they were broken down into a minute *division of labor*. From his observations of the production process—possibly the world's first time and motion study—he was able to conclude that:

> a workman not educated to this business...nor acquainted with the use of the machinery employed in it...could scarce, perhaps, with his utmost industry, make one pin in a day, and certainly could not make twenty.

However, Smith noted, if the process were subdivided into rigorously separate tasks, whereupon:

> one man draws out the wire, another straights it, a third cuts it, a fourth points it, a fifth grinds it at the top for receiving, the head; to make the head requires two or three distinct operations; to put it on is a peculiar business, to whiten the pins is another; it is even a trade by itself to put them into the paper; and the important business of making a pin is, in this manner, divided into about eighteen distinct operations, which, in some manufactories, are all performed by distinct hands, though in others the same man will sometimes perform two or three of them. I have seen a small manufactory of this kind where ten men only were employed, and...could make among them upwards of forty-eight thousand pins in a day. (1776/1965: 69)

As can easily be seen, astronomical levels of productivity may be achieved through the subdivision of labor on a work task. Clock time was central to this process. To keep the production process flowing, the individual worker would have to synchronize his or her actions with other members of the workforce to make sure "bottlenecks" did not occur in the work sequence. Workers, consequently, would have so much "time" to complete a task, and were, for the duration of the workday, driven by this precise metering. The effect of this process is the production of "economies of scale," and the essence of economies of scale, in other words, what they in fact do, is to save *time*. Saving time means the manufacturer can produce more for less and is thus able to sell more cheaply. Enabling one to sell more cheaply, as any businessperson will attest to, is close to the essence of capitalism, and the very essence of capitalism is *competition*. Clock time is thus central to the processes of capitalism, the comprehensively dominant social-organization mode in modern society. Without precise clock time metering being applied to the production process, a division of labor is not possible; economies of scale are unrealizable, cheap production of commodities impossible, and the production of exchange values illogical.

The metering of the clock regulates the production process as well, and the forms it takes today began to emerge around the beginning of the Industrial Revolution. The task-oriented workday of the peasant, a mode that had endured for a thousand years and more, was now organized into rigid and unalterable time-bound shifts: day, afternoon and night. "Clocking on" and "clocking off" framed the individual workday. Timesheets organized the tasks and the time to be taken for them, which, in turn, organized the routines of the individual, the group and the factory. So deeply did time become part of the production process that the famous statement by Benjamin Franklin in his 1748 exhortation to a young tradesman that "time is money" is at once both profound truism and shallow cliché. What Franklin meant, of course, speaking as an entrepreneur-inventor, is that time is the *employer's* time and money is the *employer's* money. Employers, then and now, buy labor on a time basis. Today, hundreds of millions of employees are still paid hourly; too many, still, are hired on a daily or weekly basis; and increasingly more are engaged to the time-length of an individual contract. As Thompson noted:

> This [time] measurement embodies a simple relationship. Those who are employed experience a distinction between their employer's time and their "own" time. And the employer must use the time of his labour, and see it is not wasted: not the tasks but the value of time when reduced to money is dominant. Time is now currency: it is not passed but spent. (1967: 61)

The importance of time to the production process, to economies of scale, to the control and disciplining of the workforce was not fully appreciated until Frederick W. Taylor's *The Principles of Scientific Management* was published in 1911. Taylor's ideas on the work process and the human participation in it revolutionized, once again, the dramatic productivity gains based on the division of labor that had so enthused Adam Smith. "Maximum efficiency" was the aim of the "scientific management" of the work process, and this was to be achieved by means of the soon-to-be infamous time and motion studies. This meant minute analysis of the physical dynamics of any particular work task and then breaking it down into the most efficient manner of completion. It meant the monitoring and timing of every detail involved in the task and identifying and eliminating any "inefficient" or "unnecessary" movements. The clock and technical calculation were the drivers of these efficiencies and not human "haphazard or rule-of-thumb" attitudes to work tasks (Taylor, 1911:100). Taylor's book was tremendously influential throughout the USA, where it was first published, and across the world. Stalin and Hitler built industrialized and efficient terror-economies from it. Thousands of businesses produced thousands of manuals based on time and motion studies to stipulate how newly "Taylorized" workers would do their old jobs more "efficiently." One manual discussed the case where a businessman complained about how slowly his secretary worked. He asked, "How many times a minute should she be able to open and close a file drawer?" The manual's answer would be "Exactly twenty-five times." Times for other open-and-close operations would be 0.04 minutes for opening or closing a folder and 0.026 minutes for opening a standard center desk drawer. Taylor and Taylorism, as Gleick (1999:215) has noted, "is by its nature brutal, stripping craftsmen of autonomy, over-riding what might have been a more natural, variable work rhythm." This was exactly its purpose: to make no distinction between the machine and its operator, or between the machine and society. Such minuscule metering of their physical movements was expected to meet with worker resistance and it did, right up until the 1970s and 1980s, when it was eclipsed by new management theories that we shall discuss below. Resistance was mitigated, however, by the application of another of Taylor's "scientific principles": the "incentive system." Timing and planning, again, are central. Workers were given a certain task, told how to do it and were given a certain timeframe in order to complete it. If they succeeded, they were rewarded with a precalculated bonus. Under Taylorism, workers not only worked to the rhythms of the clock through the actual performance of the job, but they were set on a daily race to try to *beat the clock* to qualify for a bonus.

The rapid success of Taylorism gave rise to another and even more successful "ism," Fordism, a subject we touched upon earlier. Henry Ford, the owner and founder of the Ford Motor Company, took Taylorism and its "scientific principles" from the level of management "fad" to what David Harvey called a "total way of life" that organized social and economic planning as well as industrial production (1989:135). In essence, Fordism was the principles of "scientific management" pressed into the service of "mass production," leading to supply-side driven levels of "mass consumption." Classical Fordism meant the standardization and predictability of levels of supply and demand, distribution and consumption. The planning of production and consumption led gradually, during the post-war period in the West and to a much greater extent in the Communist bloc countries, to the development of the "planned economy" based on Fordist principles. In the first half of the twentieth century, craft-based production, the mode that had still survived much of the influences of the Industrial Revolution, was being supplanted by mass production techniques in the industrial economies. Fordism and Taylorism entailed not just an increased division of labor but also increased mechanization of the labor process and the "mechanization" of the individual within that process. The trade-off for such mechanistic domination by capitalism was increased standards of living for all—or nearly all. But this trade-off was very much slanted in favor of capital, according to writers such as Herbert Marcuse (1968), who argued that Fordism, Taylorism and the "total way of life" it produced made for a "one dimensional" and alienated existence. This arid "one dimensionality" would have been impossible without the internalization of clock time into more and more facets of our lives.

In a relatively short space of time, the importance of clock time moved from the periphery of culture and society to its very core. Clock time transmogrified gradually from an abstract concept used to measure and organize the prayer times, meal times and work times of medieval monks, to the more recent pre-industrial function of ingenious mechanism that rich landowners and merchants displayed as symbols of status. It was in the process of capitalist commodity production, however, that the fantastic power of clock time was unleashed, merging with the process of commodity production to the point of being indispensable to it and at the same time becoming a commodity itself. Time became money. Time was reified from an abstract and remote concept that concerned relatively few people to a "thing," a commodity that was valuable and scarce and had its effects felt deeply and widely throughout society. However, although clock time shifted to the center of the emerging industrial society, it did not yet saturate every nook and cranny of social and cultural life. This process took much longer, and even in the most advanced industrial countries today it is still an incomplete, though ongoing, process. Clock time, from the Industrial Revolution until fairly recently, was a regular metering and measuring that, through improved communications technologies, "annihilated" or "compressed" geographic space. The result of this process is what Marshall McLuhan called the "global village." Time itself remained chronologically metered as an abstract, scientific form of measurement and organization. It was not until the information technology revolution, the so-called "third Industrial Revolution," that time began to be affected by technological development—signaling the shift from chronologic to chronoscopic time. The next chapter will outline the history and dynamics of this process and consider some of the effects that this transition is having upon cultures and societies today in the Information Age, especially upon the production of knowledge and critical thought and the impact of these new knowledges upon civil society.

Chapter Four

Globalization and the Information Technology Revolution

In his very readable if somewhat depressing book, *Faster: The Acceleration of Just About Everything* (1999), James Gleick describes what we all feel intuitively: that the pace of life is getting quicker and quicker, and there seems to be no slackening of the tempo nor end in sight. In our working lives, our leisure activities, our family life, personal relationships and cultural practices, the mantras of "efficiency" and "saving time" have infused our lives with a such sense of urgency and of "battling the clock" that we increasingly feel harried and harassed by time. As Gleick shows, in instance after instance and in realm after realm, we are pushed for time and pushed by time into an almost permanent state of frustration at our inability to devote our time properly or adequately to anything or anyone. Tasks in work, in leisure and in private life are continually in a state of incompleteness. At work we drop projects through lack of time, and in our private life we drop friends for similar reasons—and feel guilty and increasingly powerless to do anything about it. We feel guilty about the amount of time we devote to our children, our jobs and our partners. Like the proverbial hamster on the spinning wheel, we run faster and faster and never seem to get anywhere. Indeed, so fast do we run that we do not have the time to stop and consider *why* we need to run so fast or how, even, to stop. Instead we get used to it. We continue to internalize the fast pace of "modern" life and the guilt and harassment that come with it until they become a simple and seemingly unavoidable fact of living.

Gleick's book, though quite useful, is nonetheless thoroughly ahistorical—a result of the processes he describes, you might say. It gives the impression that the acceleration of life is something fairly recent, something new that affects us here and now, something dimly to do with the effects of computerization. Implicit in his argument is that previous generations enjoyed life at a rather more relaxed, comfortable and "natural" pace. In fact, the perception that life seems to get faster and faster has a rather long prehistory. At least since the beginnings of the Industrial Revolution, when clock time began to take hold in a powerful way, people have been speculating upon the cause and effects of the perceived "acceleration" of life. In the 1860s physician and specialist in the treatment of mental disorders, George Miller Beard, described an effect of the acceleration of life as "neurasthenia," or nervous exhaustion. Discussing his work in *The Female Malady* (1985:137), Elaine Showalter noted that neurasthenia was:

> originally...described as "American nervousness" by the neurologist George Miller Beard [who] saw a significant correlation between modern social organization and nervous illness. A deficiency in nervous energy was the price exacted by industrialized urban societies, competitive business and social environments, and the luxuries, vices, and excesses of modern life. Five characteristic features of nineteenth-century progress—the periodical press, steam power, the telegraph, the sciences, and especially the increased mental activity of women—could be held to blame for the sapping of American nervous strength.

Beard's theories quickly fell out of favor, but he was typical of the thinkers of the period who realized that industrialization was in some way having a profound effect upon society, even if they were not able to articulate the process clearly. This realization was not simply in the sociological sense of the shift from agriculture to industrialism or from the rural life to the formation of mass urban societies but also, from the concomitant effect of these profound changes upon our psyche, our ways of being and of seeing. The processes of industrialization produced a huge social, cultural and political flowering to an extent never experienced before in human history. Many new and deeply radical ways of thinking about ourselves and about society emerged at this time and in their varying ways, and to varying effect, inserted themselves into our collective consciousness. Their effects reverberate and resonate still. The ideologies of socialism, individualism, liberalism, the nihilism of Nietzsche or the anarchism of Kropotkin radicalized minds and energized whole societies. For the majority of these thinkers, to whom we can add Karl Marx, this shift from the myriad irregular temporalities that permeated and interpenetrated social life to the chronologic metering of the clock was something to be welcomed. For Marx, in particular, it meant an end to feudal backwardness and religious superstition. The Industrial Revolution, technology, clock time discipline and the capitalist order it sustained were, he argued, a necessary "stage" in the progress of humanity—progress leading inexorably toward communism.

Marx saw that particular technologies had particular effects upon social and economic organization. The major technologies of the Industrial Revolution, such as the steam engine and the telegraph, had twofold dynamics: they were imbued with the meter of the clock and they annihilated space. Space and time merged in the dynamics of modernity, and modernity itself represented, above all, a linear clock-based temporality that was pushing forwards, bringing the masses with it, ordering and organizing humanity upon a rational plane onwards and upwards toward the future Nirvana. In what must count as an early description of what we now call "globalization," Marx wrote in the 1848 *Communist Manifesto* that:

> [national industries] are dislodged by new industries...that no longer work up indigenous raw material, but raw material drawn from the remotest zones; industries whose products are consumed, not only at home but in every quarter of the globe. In place of old wants, satisfied by the productions of the country, we find new wants, requiring for their satisfaction the products of distant lands and climes. In place of the old local and national seclusion and self-sufficiency, we have intercourse in every direction, universal interdependence of nations.

It is important to keep in mind that modernity, the Industrial Revolution, its technologies, its social-political organizations and the developing worldview of individuals in a post-feudal world were underpinned by chronologic time. Globalization, as Marx describes it in nineteenth-century parlance, had its own temporality—a temporality unambiguously and unerringly based on the clock. Economic competition kept the clock ticking, but the *meter* of the clock remained the same until the 1970s and the onset of the crises of Fordism and of capitalism. Chronologic time ruled the first Industrial Revolution that was based upon steam power and ruled, too, the second Industrial Revolution that was energized by oil and driven by Fordism and the technologies of the machine. Today we are at the beginning of the third Industrial Revolution, and this, underpinned as it is by information technologies, is *qualitatively* different from the previous two. Information technologies not only compress space through the enhanced forms of communication they make possible, they also compress *time*.

It was only fairly recently and aided by computers, of course, that particle physicists and computer scientists were able to slice up the second-duration of chronologic time and separate it out into mind-bogglingly short durational segments. Humans have difficulty in distinguishing past from present in events that occur in a normal second-to-second sequential timeframe. However, when we get down to tenths or hundredths of a second as in, say, the differences between top 100-meter sprinters or swimmers, we need the aid of slow-motion cameras and hi-tech digital timepieces. Today computers are able to break down the second into nanoseconds—*one billionth* of a second, or what would appear to be the final mathematical abstraction of time. At these speeds, not only is it impossible to distinguish past from present or present from future, but also time itself, as a cognitive process, is suspended. Like the super-fast pictures of a bullet piercing a balloon or a droplet of milk creating its coronal splash, life and processes gauged in microseconds are suspended in time, as are our perceptions of it. Like the workings of the atomic clock, this super-fine grading of time fractions down to nanoseconds or even picoseconds (one trillionth of a second) would be fairly abstruse science, concerning only the community of physics boffins, were it not for the ubiquity and interconnectivity of computing.

It is the ubiquity of information technologies and their ability to spread across and transform almost all industries, production processes—and our relationship to these, is what makes the third Industrial Revolution truly a revolution. The Internet and interconnectivity are central to these processes. As Mark Weiser and John Seely Brown (1997:5) put it:

> Today the Internet is carrying us through an era of widespread distributed computing toward the relationship of ubiquitous computing, characterized by deeply embedding computation in the world.

It is this transformation, the "embedding [of] computation in the world," a transformation that is taking place here and now, which is the essence of the shift from chronologic to chronoscopic time. In this new chronoscopic order based on ICTs, the instantaneity of real-time begins to replace the older, slower, temporal duration of clock time. Moreover, ubiquitous computing in all its forms, such as the Internet, cell-phones, pagers, laptops, PDAs, text-messaging and so forth, are combining to create an *information ecology*, an interconnected digital environment that is annihilating both time and space in equal measure. The "here and now" becomes the constant temporal duration in the chronoscopic ordering of culture and society, and our individual and collective sense of past and present and of possible futures is diminishing as ubiquitous computing embeds itself deeper and deeper. Again, like the hamster on the spinning wheel, we are having trouble discerning this profound transformation, in large part because we are moving faster and faster in this real-time environment. As we shall see, this real-time information ecology that we inhabit more and more will have profound consequences for the ways in which knowledge is produced and how we make sense of the world. As Ron Purser (2000:5) notes:

> The shift from chronological to chronoscopic time involves a radical change in temporal orientation, and the very means by which we make sense of our lives. Chronoscopic time signals an intense compression. The extensive time of history, chronology and narrative sequence implodes into a concern and fixation with the real-time instant. What used to comprise a narrative history—sense making based on knowledge of the past, present and future—contracts into the buzz of the flickering present.

Temporal sense-making requires a certain intellectual facility, one that helps us to avoid simply parroting what we hear and mimicking what we see: the ability to think critically about the world and how it presents itself to us. This process not only creates new ways of seeing and understanding, but also—by a constant testing of ideas through debate and more critical thought—*creates new forms of knowledge*. The inculcation of this intellectual faculty and the concomitant processes of new knowledge production *take time*.

Modernity can be said to have undergone a qualitative change for the first time since Henry Ford began rolling Model T cars off the production line in 1914, and indeed a good deal of it may be said to be positive. Innovations in computerization and automation have created new jobs and new processes, eliminating at the same time those repetitive and labor-intensive tasks that made countless occupations both tedious and alienating. The ICT revolution has unquestionable public benefits, too. One only has to think of the advances made in medicine. Computers have revolutionized the sciences of pharmacology and biotechnology, producing powerful new drugs and genetic techniques that combat many diseases once thought incurable. The mapping of the human genome and all that it may hold in prospect for humanity would have been impossible without the use of supercomputers. Interconnectivity has been indispensable to these processes. Without networks of scientists and engineers working together on common projects, swapping information, testing, experimenting and refining procedures and theories, the applications and successes of computing power on its own would have been limited.

However, the drawbacks of this process are significant and over the long run may eclipse the benefits. Living in a constant present within the information ecology may be conducive to being part of a "smart" economy. We may learn to "seize the moment" of commercial opportunity, to capitalize on a new product or service through the Internet, to react with lightning speed to stock market fluctuations, or to "fast-track" that career-enhancing MBA at a "virtual university" whilst utilizing our time maximally in business, in family life, in relationships and so on. Nevertheless, my main contention is that being increasingly unable to think reflexively and critically about ourselves, or about our jobs—what did I *really* learn in that expensive MBA?—or about the information that daily enters into our lives through a lack of time will mean living in a shallower society. Ultimately this may well lead to the evolution of a society where rat cunning and quick reaction times are valorized over more considered opinion, debate and the evaluation of risk management strategies in our personal lives and in the organization of society. It will be a society that is even more one dimensional, and a society and the individuals and groups within it that will be more open to forms of control by institutions and organizations of power that are able to capitalize upon the dynamics of the information ecology. The dynamics that go to make such outcomes possible will be discussed in Parts Two and Three.

For the present, however, to be able to understand how this situation came about we need to take time to analyze the dialectical interrelationship between globalization and the IT revolution. The two are inseparable. Each augments and reinforces the other to the point that it would be fair to say that there would be no globalization without the ICT revolution, and the ICT revolution would not have begun were it not for economic and political forces that initiated the processes of globalization. It is necessary therefore to separate both subjects and unpack their interrelationships and show how and why society and its temporal metering have shifted from chronologic to chronoscopic time.

What Is Globalization?

Notwithstanding the fact that "globalization" is a fairly recent neologism, the broad dynamics of economic globalization may be dated to at least the Industrial Revolution, as the earlier quotation from the *Communist Manifesto* would suggest. This is globalization as it is generally understood, that is, the expansion of the productive forces of capitalism, of markets, of goods and services across the spatial and physical terrain of the planet. This process has progressed in three distinct phases, which I shall outline shortly. To begin with, we need to give this definition some conceptual fine-tuning if a useful and subtle understanding of the dynamics of the process is to be arrived at.

It seems illogical to speak of globalization without giving central consideration to ideas of space and spatiality. In other words, globalization is fundamentally an economic-geographic process. It was David Harvey in *The Limits to Capital* (1983) who gave pivotal consideration to what he termed the "spatial configurations of capital." He saw that the way capital "accumulated"—that is, the capitalist's drive to amass as much capital as possible to enable him or her to invest, grow the business and compete in the marketplace—was affected by the physical geography in which this process took place. Harvey noted that the process of accumulation within space has a tendency toward "over-accumulation": a point is reached at some stage within a certain marketized geographic area where accumulation produces a surplus of capital relative to opportunities to employ that capital (1983:192). There are two "solutions": one is to devalue capital (write off unsaleable assets, close plant, cut production, sack workers, etc.); the other is to resort to what Harvey terms the "spatial fix." This is to do what capitalism is inherently prone to do if it is to survive. It has to expand into fresh geographic areas where investment opportunities can be exploited in the creation of new markets, to expand in the acquisition of cheaper (or new) raw materials, and to expand to create the opportunities to exploit fresh sources of cheap labor.

Central to the periodic bursts of generalized "spatial fix" geographic expansions are communications and production technologies: the innovative use of existing technology or the introduction of revolutionary new technologies to allow this expansion to happen. In the first Industrial Revolution, the primary modes of expansion in communications and production technologies were through the steam engine, railways, steamships, telegraph and so on. These compressed space, making the "problem" of expansion feasible and, for many, extremely profitable. However, as Harvey also points out, capitalism also has a paradoxical tendency to become "fixed" in space through the fixed nature of plant and machinery, of markets and consumers. What this means is that over time a frenetic burst of expansion (such in the late nineteenth-century phase of imperialism) begins to mature and stagnate, and "over-accumulation" becomes a systemic problem once more. The world depression of the late 1890s, for example, may be seen as an outcome of this tendency.

We have discussed already the motive force behind the second Industrial Revolution—that of Fordism—and how this radical new mode of production allowed capitalism to expand and grow to unparalleled levels. In the first decades of the twentieth century, goods and services could be made cheaper and on economies of scale that were unimaginable only a generation previously. Mass production and mass consumption spurred many new industries and boosted many existing ones. A frenzy of speculation plunged the world economy into depression in the 1930s, and led ultimately to war by the end of that decade. It was after the war that Fordism reached its social, political and organizational zenith. This period (1945–1973) was the so-called "golden age" of expansion almost everywhere and prosperity for almost everyone. By 1970 the industrial output of the major capitalist economies was 180 per cent higher than in 1950, and more was produced in that last quarter of a century than in the previous three. Such sustained growth meant that output doubled every fifteen or so years and, with annual population growth at 1 per cent, each successive generation might expect to be at least twice as well off as its parents (Armstrong et al.,1984:167–8). This was the period of Fordism as a "way of life," where people could expect steady and well-paid jobs and to be in them for many years. Government could expect to be able to plan and grow the economy, providing increasing services such as welfare and health care, and social blights such as unemployment and poverty were inexorably being eradicated.

This phase of globalization came to an abrupt end in the early 1970s. Once more the problem of over-accumulation began to drag down rates of growth and expansion. The Fordist mode of production quickly began to mature and then stagnate, reaching a point of sharp economic slump and then global economic crisis in 1973. As we shall see shortly, the political and economic resolution to the crises of the 1970s led directly to the latest phase of globalization and to the third Industrial Revolution: the revolution based upon and sustained by *information*.

Why Did the ICT Revolution Happen?

It is no coincidence that the ICT revolution began when it did: around the end of the 1970s when the global economy was just beginning to show tentative signs of recovery after the slump of the middle of the decade. I said in the previous paragraph that the resolution to the global crises was both political and economic. It is important to keep this in mind if we wish to understand the nature of the current phase of globalization and the ICT revolution that underpins it. The severity of the economic crisis in the Western industrialized economies—and in the English-speaking economies in particular—provoked what we can see in retrospect was a *political revolution*: the rise of the New Right as a political movement and neoliberalism as its economic philosophy. The intellectual impetus for the New Right emerged out of the conservative think tanks in the English-speaking world such as the Heritage Foundation in the USA that was set up in 1973, and the British Centre for Policy Studies that was founded by Margaret Thatcher and Keith Joseph in 1974. Their aims were spelled out clearly. The Heritage Foundation's mission statement announces that its purpose is to "promote public policies based on the principles of free enterprise, limited government, individual freedom." Similarly, the Centre for Policy Studies "bases all its policy proposals on a set of core principles, including the value of free markets, the importance of individual choice and responsibility, and the concepts of duty...individualism and liberty." The intellectual lineage of these ideas comes more or less directly from the traditions of English Liberalism going back at least to Adam Smith.

The 1970s brand of (neo) liberalism saw the twin factors of the overweening state and overly powerful organized labor as the crux of the problem, with a two-pronged political and economic attack required to address it. Crucial to getting the state "off the back" of free-market entrepreneurs, and cutting the overbearing and meddlesome unions "down to size," was having politicians in power who believed this was the right thing to do—and would be prepared to do whatever was necessary to achieve it. Enter the Reagan-Thatcher era. The administrations of Ronald Reagan in the USA and Margaret Thatcher in Britain blazed the trail and set out to "restructure" their domestic economy and overhaul what was viewed by them as the coddled and inefficient "way of life" that had grown up around post-war Fordism. The discipline of the market and the "natural" logic of market forces, the neoliberals demanded, were all that were needed to drag the world economy out of economic and social crisis. Accordingly, the theories of Adam Smith and latter-day disciples such as Friedrich von Hayek and Milton Friedman were suddenly in vogue. These were applied ruthlessly to the economies of the English-speaking world and then to the rest of the world as this powerful ideological doctrine spread (Kolko, 1988). As a result, the restructuring of global capitalism was soon underway.

The need for free and open "competition" was a key feature of neoliberal thinking. For there to be "fairness" in the competitive marketplace, the owners and controllers of capital needed to have a free hand in the running of their businesses, with little—or preferably, no—interference from either government or organized labor. Bosses needed to have autonomy. This meant freedom from meddling governments and unions to decide *what* is to be produced in their factories and offices, *how* it would be produced and *where* it would be produced. From the late 1970s until today, New Right governments of Reagan and Thatcher and their successors in the wider Anglo-Saxon world have gone to extraordinary lengths to vouchsafe this "inalienable" right to capitalists. Notoriously, Thatcher referred to coalminers who went on strike during 1984–85 to defend their jobs, as "the enemy within," and she used all the powers of the state, from the police to MI5 (the British equivalent of the CIA) and the Special Branch, to undermine the industrial action and ultimately defeat the miners (Milne, 1994). "Flexibilization" was the neologism that was key to what was needed to be able to compete successfully: flexible labor markets, flexible capital markets and flexible production methods. The latter were construed as the right to introduce any technology or working practice that employers deemed necessary in the race to attain the all-important "competitive edge."

During the Fordist "golden age" boom from the 1940s to the 1970s in most developed economies, the tripartite alliance of "big labor, big capital and big government" would work together and compromise around issues of mutual interest such as wages, prices and productivity. This "give and take" process (one, incidentally, that was historically unique within capitalism) was easily accommodated in the climate of rising profits, rising wages and swelling treasury coffers. This is not to suggest that computers and automation were not being used in production and administrative tasks in many industries. The Fiat automobile company had introduced rudimentary robots into their production lines as early as 1972 to do repetitive tasks such as spot-welding and spray-painting. Similarly, the computerization of certain work tasks had been in operation since the mid-1950s in large insurance companies and government bureaucracies for the processing of invoices, census data, application forms and a great number of other routine administrative jobs. However, given the power of organized labor, the "reasonableness" of employers and the wish by governments to keep unemployment down, the introduction of new information-based technologies was very slow and implemented in such a way as to avoid "displacing" labor. Again, during the period of boom-for-all this was accommodated without too much difficulty.

In the new era of crisis that was the late 1970s and early 1980s, neoliberals argued increasingly the need to give as complete autonomy as possible to employers in the running of their own businesses in the quest for competitiveness, flexibility and profitability. Ipso facto this meant that the traditional prerogatives of organized labor were to be overlooked in the introduction of new technologies. Indeed, far from being consulted in this brave new world, labor unions were more often than not crushed if they attempted to resist sweeping change. A series of set-piece labor-employer battles during the early 1980s such as the air controllers' strike in the USA and the miners' strike in Britain indicated where neoliberal government stood on these matters—and it was squarely behind the employer. A vicious cycle began to set in. The forcible introduction of new technologies—with the now-explicit aim of "displacing" labor to save costs or to modernize production methods or to close down altogether and move offshore where costs were less and unions non-existent—sent unemployment in most of the OECD countries to 1930s levels. This in turn severely weakened union power and emboldened employers still further to reassert control over the whole of the production process, relying more and more upon hi-tech solutions in the race for the "competitive edge." The shift toward computerization and automation was swift and remarkable, to the extent that during the early part of the 1980s more of the investment dollar in the USA went into computer and related hi-tech equipment than into traditional labor-intensive machinery: an historically unprecedented investment shift (Kolko, 1988: 66).

"Supervening Necessity" and Technological Convergence

Given the level of hype that surrounds the information technology revolution, one could be forgiven for thinking that many hi-tech applications such as the computer, the satellite, the facsimile machine or even the Internet are relatively recent innovations. In fact, these technologies or their direct antecedents are rather old, and in some cases very old indeed. The first fax (a photographic image) was transmitted from Munich to Berlin in 1907. The laser technique used to send light and data down a fiber optic cable was developed in 1960, based on technical antecedents created by Alexander Graham Bell in 1880. Satellite communication technology, a component of what Manuel Castells has dubbed the "network society," was developed and put into operation with the launch by the Soviet Union of *Sputnik I* in 1957. The prehistory of today's computer is even longer. The quest for a workable mechanical calculator, a machine that would take the boredom out of mathematical computation, began at least as early as the 1830s with the work of Charles Babbage and his "analytical engine," a project that was never realized. However, Babbage's work and vision were tremendously influential and contributed to the emergence of working mechanical computers during the 1930s. It is well known that the analogue telephone dates from the late nineteenth century. Less well known is the fact that the "hi-tech" cellular telephones that so suddenly saturated the marketplace in the late 1990s represented a largely incremental step in a process of technological development that had been underway since the 1920s. And the Internet (basically a network of interconnected computers), a technology that has been subjected to extraordinary levels of hype since its general uptake in 1995, was conceived in theory in the 1940s by Vannevar Bush. A working reality and direct precursor of the Internet came in the 1960s in the form of the military-based communications technologies used by the Advanced Research Projects Agency (ARPA) and evolved into the university-based ARPA-Net communications system in the early 1970s. The first email was sent in 1972—as was the first piece of spam-mail.

As we have seen, the period from the 1940s until the 1970s, the period when many of these technologies had been developed, was also the period of the Fordist "golden age" boom. In this climate of profit and prosperity for all, the "social contract" between capital, labor and government created a social, political and cultural atmosphere that was almost akin to complacency. Highly influential books began to appear, such as Daniel Bell's *The End of Ideology* (1962), which reflected the sense that the big ideological and political battles were over: consensus ruled. The Cold War against the Soviet Union was the only ideological and political threat and created its own military-industrial dynamic with regard to technological innovation. ARPANet was developed out of the profound shock that the launching of *Sputnik* caused the US military and political elite. In the main, however, the pace of overall technological innovation during the "golden age" boom was relatively sluggish and tended to follow discrete developmental trajectories. In the climate of plenty, things just seemed to tick over nicely, and there did not yet exist what Brian Winston has termed a "supervening necessity" to bring forth radical technological, political and economic changes (1998:147). However, the severity of the mid-1970s global recession provided just the right context for such a "supervening necessity." The post-war years of complacency came abruptly to a halt, and so deep and severe was the jolt of economic crisis that increasing numbers of neoliberal politicians, industry lobby groups and neoliberal think tanks demanded far-reaching measures. The political revolution of the New Right and neoliberalism constructed the environment of new economic realities and—as our political and business leaders continually reminded us—these realities were harsh realities. Consensus was out and ruthless economic competition and social conflict (dynamics that had been in relative slumber since the 1920s and 1930s) were back in.

In the new climate of what seemed to be perpetual crises in the late 1970s and early 1980s, the mantras of "efficiency," "restructuring," "competition" and above all the manager's "right to manage" meant that all bets were off. The Fordist "social contract" was unilaterally torn up by big business and supported in this by neoliberal governments. The stage was now set for "technological convergence." How did this work in practice? The New Right emphasized the need to expunge the "rigidities" of Fordism through economic restructuring. Neoliberal economic theory defined increased productivity as the key to competitiveness and hence profitability in the new post-Fordist world of the 1980s and 1990s. The winners in this new "global marketplace" would, therefore, be those companies whose use of capital investment was as unrestricted as possible and was put to the most productive use possible. The enormous potential for productivity increases through information technologies had long been recognized (Weiner, 1968), and it was in the new climate of intensified economic competition that the ICT revolution really got underway. Perspicacious (and ruthless) producers quick to recognize the productivity advantages of information technologies began to make tremendous profits as their competitive advantage paid off. ICTs began to be seen as *the* path to greater productivity and profit, and so the diffusion of ICTs across as much of the production process as possible became a priority for almost every industry (Hassan, 2000). Moreover, information technologies were quickly found to have "enabling capabilities," which meant easy diffusion and applicability across almost any industry and almost any process. Whereas previous revolutions in productivity tended to be confined to particular industries, with ripple effects spreading relatively slowly outwards, computers and automation could—and did—quickly revolutionize the design, analysis and production of everything from automobiles and DNA mapping to architectural design and publishing.

The all-encompassing nature of information technologies meant that it began to affect hitherto discrete information communication technologies, drawing them together in a process of "convergence." Satellite, voice telephony, fiber optics and computer technology converged rapidly into a meta-communication technology that further supercharged the productive and diffusionary capabilities of ICTs. The quantum technological jump achieved through the convergence process also rebounded back upon the once-discrete technologies themselves, to digitize and supercharge them in their own right. Take the case of humble telephony. The processes of convergence have digitized this and radically changed its technological base, spawning, amongst other things, highly flexible and powerful wireless communications. In turn, this dragged cell-phone technology out of its deep sleep, propelling it into one of the fastest-growing and dynamic industries within the ICT sector. The take-up rates of these devices, together with convergence-led applications such as Wireless Access Protocol (WAP) and 3G enabled communicators, deepen still further the density of networks and the interconnectivity of people in the social, market and private spheres.

Convergence = Networks = Globalization

It was this process of technological convergence that made possible the emergence of Castells' "network society" (1999). Of course, to an extent, human societies have always established networks of some sort or another. And at least since the so-called Age of Discovery of the fifteenth and sixteenth centuries when European explorers opened up the East and West in the quest for spices, slaves and gold, societies began to develop *systematic* networks. These early networks were based on trade routes and on markets for tradable goods. Their sophistication, efficiency and density reflected both the mode of production (agriculture, industrial) and the concomitant level of technological development prevalent at any particular time. Competition (economic and military), from the very beginning, has always been at the center of network building—or destroying. From the fifteenth century onwards, when the leading European countries such as Holland, Portugal, France, Britain and Spain began to fashion their own trade routes, "interconnectivity" *between* their respective networks was minimal. Networks of diplomacy aside, when contact came it was usually of a violent and military nature, in the form of the plunder of each others' ships, raiding of trading posts and so on in a non-systematic and opportunistic fashion. Temporally speaking, these networks reflected the rhythms of each individual country and their own political, social and economic make-up. Accordingly, there was a large degree of asynchronicity between each country's networks, with a trading network able to remain undisturbed for many years, having little or no contact with the trading networks of other countries.

During the frenzied activity of the Industrial Revolution, social, military and economic networks "naturally" became much denser, more organized and more systematic—reflecting, as they did, the needs of industrial production and distribution. Capitalism needed regularization and predictability. Supplies of raw materials, distribution chains, labor supply and access to sea routes all needed levels of planning and organization that entailed new forms of industrial and national bureaucracy and a deepening interconnectivity between rival networks. It is something of a contradiction within capitalism that intense economic rivalry goes hand in hand with increasing levels of cooperation. Capitalists need to cooperate *and* compete. Rival countries and rival companies began to reach out and develop relationships with each other, relationships of "mutual benefit" that, in turn, deepened and widened the networks of society in general. The changing perceptions of space and time were central to the development of these "modern" networked societies. Around this time, the metering of the clock began to come into its own, synchronizing people, industries, societies and economies with its particular rhythm. The world began to be perceived and comprehensible as one place for the first time. Wars in faraway places could have an effect at home, causing panic and/or bouts of jingoism. A crop failure in Virginia, say, could be felt in the wallet of a pipe-smoker in Berlin or the tobacco stockholder in London. From around the last quarter of the nineteenth century until the beginning of the First World War, a profound transformation took place in the creation of the networked society of modernity. The development of new communications technologies emerged out of the new evolving networked society and, in turn, contributed to its ongoing development. As Stephen Kern (1983:1) put it:

> From [1880 to 1914]...a series of sweeping changes in technology and culture created distinctive new modes of thinking about and experiencing time and space. Technological innovations including the telephone, wireless telegraph, x-ray, cinema, bicycle, automobile, and airplane established the material foundation for this orientation...the result was a transformation in the dimension of life and thought.

These were exhilarating times to be sure, but in retrospect we can see that technological diffusion was relatively slow, and their developmental trajectories remained fairly discrete. After the initial burst of innovation, things began to settle down to a particular rhythm and pace—set by clock time and rhythms of the Fordist mode of production. Fordism began to grow into itself as a mode of production and set the temporality and rhythm of life that would dominate until the 1970s. Communications networks during this period certainly grew steadily, but costs were high and affordable only to the relatively wealthy and to large corporations. Global networks were fairly comprehensive but with a low level of density. For example, direct telephone dialing from the USA to Britain was available to most subscribers around 1971, but the cost of a call from New York to London was in relative terms the same as it had been in 1940, which was more than most people could easily pay.

Nevertheless, within this overarching industrial rhythm there existed still a multiplicity of irregular temporalities that interspersed and interpenetrated the lives of people in society. Low-density communications networks meant that there were many interstices of irregular rhythms that were marked by their asynchronicity, and most individuals were still able to demarcate work-time from leisure-time fairly unproblematically. Hobbies, pastimes and the general patterns of cultural practice within communities could still remain wholly outside the domain of Fordism. It was argued earlier that Fordism had become a "way of life," and it was, but there nevertheless still existed places (spaces) where people could pursue other interests and construct other ways of being and seeing. These may be tangentially linked to the realm of Fordist life, or closely linked to it—or not linked to it at all. Social, cultural and economic life did not yet constitute an information environment.

The rise of the networked society was directly attributable to the dynamics of technological convergence that we have just discussed. Central to Castells' (1999:30) theory of the "rise of the network society" is the claim that:

> Around [this] nucleus of information technologies...a constellation of major technological breakthroughs has been taking place in the last two decades of the twentieth century in advanced materials, in energy sources, in medical applications, in manufacturing techniques (current or potential, such as in nanotechnology), and in transportation technology among others....*The current process of technological transformation expands exponentially to create an interface between technological fields through common digital language* in which information is generated, stored, retrieved, processed and transmitted....We live in a world that...has become digital. (italics added)

Convergence made this process possible, creating a super-powerful technology based upon the generation, storage, retrieval and processing of information. As I indicated earlier, ICTs are "enabling" technologies that can transform almost every industry and have the prolific ability to create many new ones—centered around, and dependent upon, information. Information technologies based upon computerization and digitization (binary codes) are ideally suited to the building of networks as they contain what Castells calls a "networking logic," that is, they can "talk" to each other (1999:61). These growing networks are constantly connecting through wireless, satellite, telephony and fiber optics—and these vectors all connect with each other, to build a "network of networks." The foremost vector, the preeminent "network of networks," is the Internet. Through networking, information feeds information and information creates more information, not all of it useful and good, to be sure, but it circulates in a cumulative feedback loop that has been growing exponentially in size and density over the last twenty years.

Castells sets great store by what he terms "the pervasiveness of the effects of new [information] technologies" (1999:61):

> Because information is an integral part of all human activity, all processes of our individual and collective existence are directly shaped (although certainly not determined) by the new technological medium.

Here Castells seems to be arguing that due to an "integral" human trait (the centrality of information to our lives), we simply "soak up" information technologies: ipso facto information technologies are beginning to pervade every nook and cranny of culture and society. I think he is only partly right. Certainly, information is an "integral part of all human activity," but I do not think that we "naturally" allow it to pervade our lives. To understand why information technologies pervade our lives we need to go back to our discussion on *why* the ICT revolution happened. It did not drop out of the sky—revolutions never do; they are explosive articulations of social, political and economic conflict that have reached crisis point. It was set in motion through a political and economic reaction to the crises of the 1970s. The "resolution" to the crises was the New Right, neoliberal project to "unshackle" business and capital from the dead weights of Fordism and organized labor. Convergence and the information technology revolution were an unintended consequence of capitalism's collective efforts to make itself profitable through restructuring and production efficiencies and labor-saving through automation and computerization. The information technology revolution was a *business* revolution that organized labor, unorganized individuals and governments across the world have been trying to keep up with and come to terms with ever since.

Beginning around the early 1980s, the enormous potential and utility of ICTs in the quest for profit and economies of scale dawned on the general business community: and what Theodore Roszak (1986) called "the cult of information" began to dominate attitudes toward it. The period of economic crisis ended and economies began to grow—as did businesses. In fact, businesses went *global*, with "globalization" really coming into its own. It is estimated that the number of transnational corporations rose from 7,000 in 1969 to 37,000 in 1994 (*Economist*, 30 July 1994). As economies grew and the number of business grew alongside them, markets inevitably became tighter and more competitive. Businesses responded to difficult market conditions by more cost cutting, more automation, more economies of scale and by expanding the business into new markets, developing new markets through new products and services, and taking over or merging with competitors to achieve this same end. This process of expansion is one element in the dynamic of globalization, what may be termed "outward globalization." However, the colonizing logic of globalization also began to turn "inward" into realms of culture and society that hitherto had been free (or relatively free) from the dynamics of the market. Through this powerful neoliberal globalization/ICT revolution nexus, the demarcation lines between the private and the public realms began to blur; business networks and market networks began to impinge and connect within every nook and cranny of social and cultural life. The accoutrements of the Information Age such as the Internet, the cell-phone, Web TV, ICQ, text messaging, email and so on increasingly plug us into "networks of networks." These, in turn, connect us to the rhythms and economic forces of globalization. And globalization rebounds back upon us both as individuals and as members of societies. Interconnectivity affects us through flexible working, short-term contracts, the vagaries of the stock market, and the commercialization and deregulation of the very core of the public services that had evolved over the last one hundred years, such as public education, transport, health, welfare and the prison system.

Ubiquity and pervasiveness are indeed central to the information technology revolution, and globalization and interconnectivity are daily making the networked society denser and denser (I shall discuss this in more detail shortly). This growing thicket of interconnectedness is creating the information ecology that is at the center of my argument about the production of knowledge as we move through the twenty-first century. This information ecology pervades, increasingly, much of what we do in our private lives and in our public working lives. As this information ecology becomes denser through increased interconnectivity across the growing range of networkable devices, there will be fewer spaces, fewer interstices where a non-market, non-commercialized existence can grow and thrive, spaces where alternatives can be conceived and practiced. The information ecology emerges out of the imperatives of the New Economy, and interconnectivity within this New Economy synchronizes neatly with the needs of business, creating connected markets and distribution networks that span the globe and represent hundreds of millions of people. The global digital "networks of networks" transcend time and space. Geographic space is negated and transformed into virtual space, and the age-old metering of clock time is reduced to real-time that is gauged in nanoseconds. In short, the globalization/ICT nexus represents the shift from chronologic to chronoscopic time.

For some such as Nicholas Negroponte, founder of MIT Media Lab and *Wired* magazine guru, the shift to a digital economy is a good thing. Indeed, his mission is to blur the boundaries between "bits and atoms" where the physical and the digital come together, creating digital people who inhabit a digital world (Negroponte,1995). Somewhat less effusive perhaps, certainly less visionary, but no less enthusiastic are those many people in government, the media and industry around the world who, like

Bill Gates (1995), foresee a world of "friction free capitalism." Here it is predicted that the boom-bust cycle that has tormented economic development for two hundred years will be erased, and prosperity, diversity and freedom will issue forth. ICTs will, it is claimed, create a "knowledge nation," or a "smart economy." Too few, however, have pondered (or have the time to ponder) the effects of the shift to a digital chronoscopic order, where real-time reigns and perception changes. This is not a technological determinist position I am taking, a simplistic argument whereby a certain technology "makes" us act in a certain way. What we face is something unprecedented in modern society: the creation of a whole environment, an ecology based upon ICTs, something that is reaching into every part of culture and society, networking, building networks upon networks, and running these on a real-time basis. We interact with our environments, sure, but ICTs have taken this interaction to a new temporal level based on chronoscopic metering. Within this environment we can exert some control, become smart, even change elements of our environment, such as humans have always done (for good or ill). However, living on a chronoscopic plane, we will have a degraded sense of our past and inhabit a constant present. Before developing the arguments that get to the core of this book, living and learning in a chronoscopic environment, it may be useful to set the scene with some remarks about what life in the digital ecology entails.

Chapter Five

Digital People in a Digital Ecology

It is always useful to be wary of claims made for new technologies. Usually, if one peers just beneath the surface of such claims, commercial interests lurk, hyping their product for all it is worth. Even if there are no immediate commercial interests at work, unbiased claims regarding the predicted impact of this or that technology regularly fail to live up to expectations. History is littered with the hastily dug shallow graves of allegedly world-shattering products that quickly fell into normalcy, obscurity or ridicule. Our own time, especially, is replete with claims that seemed at the time fairly plausible but soon shot embarrassingly wide of the mark. For instance, during the 1970s and early 1980s we were continually and breathlessly informed that the "paperless office" was a reality just around the corner. Gaudy adverts in glossy magazines for IBM or Honeywell depicted actors with flared pants and jaw-length sideburns relaxed and smiling next to an enormous computer, with only cool IKEA-type furniture to disrupt the smooth lines of office efficiency. The first step to this space age and productive Nirvana, of course, was to computerize the office. The personal computer would then store everything we wanted electronically, and all we had to do was call it up from the hard drive, read it on screen, and then send it back to the hard drive when we are finished. Countless millions of documents could be stored in this way, virtually, electronically, cleanly and with not a scrap of paper in sight. The unproductive and messy clutter of the office would vanish, and in its place would be streamlined efficiency and profitability. But the ICT revolution is powered fundamentally by competition, and so nothing is able to stay the same for too long. Printing your own documents on your very own new desktop printer was pitched as the next phase in convenience, the ultimate in document distribution. Hard copies were more "secure," we were told. Out went the idea of the paperless office, and so papermaking and printing on a vast scale came in. Most offices are now awash in paper, using far, far more than ever before. The same story can be told about the fax machine. According to business consultants Don Peppers and Martha Rogers in *The One to One Future*, the fax machine (a technology that dates back to 1907, remember) represented a "paradigm shift" that would result in "cataclysmic changes" in society (1993:3–10). Millions of fax machines have been sold since the mid-1980s, but the "cataclysm" just never seemed to arrive. In offices across the world, most fax machines sit silent and neglected in the corner for most of the time. They quietly gather dust or have been ruthlessly consigned to the landfill to make way for a new laser printer, or scanner, or server or whichever new networkable device is billed as the new "paradigm shift" technology or killer app.

The Internet has probably been subjected to more hype than any technology in history. Part of this is because so many businesses have so much riding on it—the media industry (the prime vector for the hype) being no exception. But the fact that so many people and businesses, from across almost every industry, have invested heavily in it suggests that something different has happened. Consultants or marketers pitch a single device with as much hype as they are able to get away with. This is understandable. In a competitive marketplace this is what they are paid for. Undoubtedly, much that is written about the Internet needs to be read skeptically, too, with a view to who is writing it, who they work for, or which organization they belong to or own. Having issued that caution, it still needs to be said that neither the fax machine, or the printer, or the scanner or the server or the cell-phone ever spawned its "gurus" or prophets or research institutes or university degree subjects or a vast literature ranging from self-serving nonsense to the thoughtful and insightful. Either right or completely wrong, almost everybody who has written about the Internet, developed a product to attach to it or invested their money in it recognized, at least at some level, that something extraordinary was afoot. This is not a singular device or contraption that would fall or stand in the marketplace—usually in short order. It isn't anything tangible that could break, or rust, or look ridiculously old-fashioned in a year or two. The Internet is almost an organic *totality* in the making. It is virtual and concrete at the same time, amorphous yet rational, non-existent and yet potentially all pervasive. The possibilities, to resort to the adman's cliché, are literally limited only by our imaginations. Central to this is the potential for limitless and incredibly dense networks that can reach into every facet of our lives, draw us into them and connect us to their real-but-virtual world(s). I am not talking simply about the Internet in its stripped-down form of a network of interconnected computers, important though this is, but in its role as "backbone" for the networks of networks, for everything that can and will, in the future, be connected to it.

It may be useful here to ground this argument in some empirical data to get an idea of the extent of what is happening and to gauge the proportions of the growing information ecology. Take the Internet, the "backbone" of this ecology. Prior to 1995 it was primarily boffin and university researcher territory. Since that time the Internet has burst onto the world scene with a rapidity that is unprecedented for a new technology. As Anthony Giddens (1999:12) observes:

> It took 40 years for radio in the United States to gain an audience of 50 million. The same number was using personal computers only 15 years after the personal computer was introduced. It needed a mere 4 years, after it was made available, for 50 million Americans to be regularly using the Internet.

The take-up rate for cell-phones has been even greater. But let us continue looking briefly at the growth of the Internet. If anything, Giddens' figure for Internet density is conservative. Methodologies for counting differ and numbers can fluctuate considerably, but Computer Industry Almanac Inc., an IT research company, conservatively put the number of Internet users in the USA at 149 million in 2002, and expects that by 2007 there will be 236 million online users in that country alone. Although the USA has the heaviest Internet-user density, other countries and other regions are catching up. Worldwide, the number of users in 2001 was 533 million and is expected to increase to over 1.4 billion by 2007 (Computer Industry Almanac, 2002). The Internet demographics company Nielsen Net Ratings (2001) estimated the number of people with Internet access across the world in 2002 to be 429 million. Internet use, access and application seem set only to grow and grow exponentially. Competition has helped the Internet become available to more and more people and is becoming more and more indispensable. Cheap computers, low connection costs and, in many cases, free access mean the processes of placing the Internet at the center of our communications environment will continue. Interconnectivity through the Internet and its growing number of peripherals has become matter of fact for many in the developed world and for a growing proportion in the developing countries, especially in North and Southeast Asia. Some appreciation of the density of interconnectivity may be appreciated if we consider that the first email was sent in 1972, a relatively short time ago. In 2001 the daily volume of email traffic was 10 billion, and this is projected to balloon out to 35 billion per day by 2005—an average of about six emails a day for every single person on earth (Moisan, 2001).

Cell-phones began to enter the market during the mid-1980s. These were big cumbersome handsets, poorly designed and expensive to buy and to use. Because they looked like Second World War field telephones growth was slow, and they remained a status symbol (and definitely not a fashion accessory/network application) for another decade or so. In 1987 there were only about one million subscribers in the USA. However, during the late 1990s in response to technological advances and the potential for huge profits for companies in the telecommunications, computer, microchip and Internet industries, cell-phone sales began to soar. In 2000 alone 412.7 million cell-phone handsets were sold worldwide; an increase of 46 per cent from the previous year (Perera, 2001). In 2000 some 52 million U.S. households, or about 51 per cent of the total, owned a cell-phone, up sharply from 32 per cent in 1999. Telecommunications analysts EMC (2002) estimated that the number of cell-phone users had topped one billion for the first time. The number of worldwide landline subscribers, the "old style" telephone, was around 800 million in 2001. Industry analysts have predicted that the number of cell-phone subscribers will surpass this by around 2005. This is an extraordinary shift. The fixed landline phone was in many ways emblematic of the Fordist period: static, predictable, functional—and *limited*. Cell-phones, however, are coming onto the market almost every week with a new connectable function or device. WAP enabled phones, where people could access limited features of the Internet, failed to impress consumers in 2000, despite the industry hype surrounding them. However, so fierce is the competition, and so great the investment, that variations of WAP that are more user-friendly such as GPS (Global Positioning Satellite) and 3G (third generation) communicators look set to take up where the WAP failure left off. The early indications are that 3G phones, which have more processing power and better graphics, will do much better. The cell-phone-Internet connection is one that simply has to be made in the eyes of industry investors, manufacturers and marketers.

Cell-phones have, in a very short time, become a sociological phenomenon as well, emblematic of the Information Age. One cannot walk in a street, board a bus or a train, enter a restaurant or a café—practically any public place—without being within earshot of a one-sided conversation taking place over a cell-phone. The same can be said for the current craze for text-messaging; youngsters, usually, focus intensely on the messages being read over their cell-phone LED screen or thumb one in themselves to someone else, somewhere else, enveloped in the information ecology and seemingly oblivious to their physical surroundings. And this very public realm of interconnectivity, one that most in the West would be familiar with, is very much the surface expression of the networked society. The roots of the digital ecology are deep within the economic processes that underpin both globalization and the information technology revolution.

E-commerce and e-business, both predicated on denser and denser levels of interconnectivity, were the industry and investor buzzwords from around 1998 onwards, spawning a rash of dotcom industries hoping to exploit a "new" niche in the New Economy. Moreover, like the Internet generally, these forms of trade have spawned hundreds (if not thousands) of books, a clutch of gurus, dozens of research institutes, and a gaggle of university degrees that major in these subjects. Internet guru and management theorist Don Tapscott has conflated these terms to construct the neologism "b-webs" or "business webs." In *Digital Capital* (2000), he views these as nothing less than representing the future of capitalism. This is essentially the "friction-free capitalism" thesis of 1995 vintage Bill Gates, where "customers [will] have more power than ever before" through increased networking with business and increased networking between businesses (2000:23). For Tapscott, it is the Internet that will coordinate and equalize in this brave new world, where transaction costs will be slashed, providing cheaper products for consumers and "exceptionally high returns on invested capital" for perspicacious and "innovative" capitalists. In "b-webs," both producer and consumer are "bathed in knowledge," providing a "synergy" resulting in "polymediation" (2000:22–3). If one manages to hack through the thicket of buzzwords and neologisms, all he is really saying is that the Internet and "b-webs" are very good for business and good for customers.

I think Tapscott may be onto something as regards business, but the jury is out in terms of benefits for customers. "Business to business" "b-webs" are far and away where most online business is done today. Forrester Research (2001) predicts continued "explosive" growth in this sector, approaching $1.3 trillion in 2003, with around 88 per cent of businesses planning or actually doing at least part of their trade online. For them, the attractiveness of "b-webs" is clear: transaction cost savings, lower administration costs, predictability in scheduling of supplies, distribution, billing, payments and so on. Consumers still need convincing, though, with growth in the "business to customer" sector much more modest, projected to be about $7-$8 billion by 2003—although this, too, is growing rapidly at 70 per cent per year. The benefits of buying online from your own home are somewhat nebulous and people recognize this. Shopping, buying in the marketplace, has been a daily occupation for many people in almost every society for at least a thousand years. The cultural patternings of this aspect of economic and social life run deep and are not so easily erased. Indeed, many goods purchased on the Internet do not represent a cost saving to the customer and are in many cases more expensive, and any initial thrill at buying a CD or book online rapidly wears off. However, the benefits for business in this online relationship are many, and consumers are being drawn into it through cajoling, enticements and, increasingly, old-fashioned economic coercion through lack of choice. We can see this most clearly in banking, where passbooks are now a thing of the past and most customers simply must have a plastic card in order to transact with the bank and with increasing numbers of vendors. Control of the production-consumption cycle is the goal for corporations in the networked society. Jeremy Rifkin (2000:102) puts it bluntly:

> Controlling the customer is the final stage in a long commercial journey marked by the increasing wresting away of both ownership and control of economic life from the hands of the masses and into the hands of corporate institutions.

Talk in the late 1990s of the New Economy in the business and mainstream media has been underpinned by subtle or explicit rhetoric regarding its *inevitability*. Like free markets, and like neoliberal globalization, the powers and benefits of the Information Age were/are presented as unavoidable, no alternative possible. The rhetoric comes to us every day, in many ways and from seemingly diverse sources. For example, Andy Grove, CEO of microchip maker Intel, asserted loftily in *Wired* magazine that "Technological change and its effects are inevitable. Stopping them is not an option" (Rossetto, 1998). You might expect a dash of self-interest here. However, when politicians (almost all of them) start talking about the inevitability and the unalloyed benefits of the Information Age, then we can gauge how deeply the ideology has taken root. Consequently, over the last ten to fifteen years, the power of "inevitability" has become a self-fulfilling prophecy.

Electronic networks are not *ipso facto* beneficent, although they are nearly always presented as such. What electronic networks *are*, undoubtedly, are new forms of social, economic, political and cultural organization. The dispersal of *power* within these new networks of networks is the acid test in terms of their levels of democracy, inclusiveness and diversity—and it has to be said the electronic networks of today's Information Age are very much *networks of domination* by global corporations. This is only to be expected because, as I have tried to argue, the motive force behind the globalization-ICT revolution has been commercial and corporate. Levels of concentration in media, in telecommunications, in software manufacturing and in microchip production—the elemental components of the Information Age—are unprecedented and continue apace. The massive mergers and acquisitions in these realms during the last decade have been breathtaking in their size and breathtaking in the fact that they are allowed to proceed. The colossal accountancy scandals in 2002 such as WorldCom simply underscored how much is at stake. Something like 40 per cent of Internet traffic used WorldCom's network at some point, and this is traffic (users) that may be taken over by a rival should WorldCom shrink or collapse. Domination, as ever in capitalist competition, is the name of the game. This domination faces millions of users every day when they switch on their computers and interface—"inevitably"—with Windows software. Lack of real control and choice underscores our experience almost anywhere we look within information networks. In news, in entertainment, in banking, in education, in commerce generally, content and delivery can be traced back to a handful of giant media, telecommunications and information technology sources. What has been hailed as a technological triumph of diversity is in reality usually a thin variation on a narrow theme. Inevitability has become a reality through the sheer force of those powerful "stakeholders" whose interests are represented in the creation of the networked society.

The so-called "change agents" of the post-Fordist world such as Wall Street investors, Internet gurus and management theorists like to describe the ICT revolution in grandiose terms. Again, their over-hyping and pushing of certain IT stocks was a major factor in the dotcom collapse of 2000. ICTs were "revolutionizing" everything, and so the "dope on the phone" was easily persuaded by the analysts to part with his or her money (Frank, 2001:110). Peppers and Rogers (1993) preferred the term "cataclysmic" to describe the nature of the changes we are going through. Such terms, though impressive and arresting, are not really useful if we are to understand the true nature of these changes. They denote something literally explosive and shattering, something that makes us turn our heads and drop our jaws in wonderment or horror. Revolutionary change is simply not like that: it is a process, an experience that may be concentrated and heightened, certainly, but not cataclysmic. Even during the height of the Russian Revolution in 1917, stormy and passionate meetings were held discussing whether or not what they were experiencing was in fact a revolution. Revolutionary change is only fully appreciable with the benefit of hindsight. As we move through the information technology revolution, our relationship to it is more one of *adoption* and *adaptation*, something we internalize as we go along in our everyday lives, something that gradually changes the way we think and operate and relate to the world and others in it. Like the introduction of the printing press, revolutionary change alters human consciousness, but the process is one of subtle internalization—not "cataclysmic" shock. I write these words on a standard PC using *Windows 2000* software, accessing the Internet periodically for all sorts of information that I can use. I correspond with colleagues through email, sending drafts and receiving comments and criticism in turn, back through the network. When I began university as a student not so long ago, I used an electric typewriter to write essays. Email and the PC were exotic and the Internet didn't exist in the mainstream. Looking back, I can't imagine how I managed to work then, so deeply has the new technology embedded itself in me. I have internalized its processes over the last few years, and it has changed my consciousness regarding the mechanics of the academic process. The trouble is I don't feel liberated or free or more efficient or productive or anything else that was promised by the manufacturers or the university bodies who decided to automate and computerize my university. What I *do* feel (when I think about it) is that I did not have a choice in the matter, and I am compelled to use these technologies for good or ill. The "acceleration of just about everything" in the shift to chronoscopic time meant that I along with most others in the developed economies was swept up in a process that was and is outside of our (individual) control.

The information technology revolution could be labeled more accurately (and more prosaically) "information technology domination by stealth." It comes to us from above with the promise of more choice, more freedom, more convenience, more efficiency and, even, more democracy—when, in fact, in many ways it is the very antithesis of these. Gene Rochlin (1997) described this form of hi-tech domination as being "trapped in the net." But it is more than this. It is more than the Internet. It is domination through denser and denser ICT networks, contributing to an acceleration of the pace of life on a chronoscopic metering. It is a revolution in the workplace where uncertainty rules and fear pervades, where interconnectivity can mean a bonus electronically deposited in your bank account one day or an email appearing in your in-box informing you are sacked the next. The logic moves inexorably toward being constantly pressed for time, finding it more and more "inconvenient" to take holidays, or devote enough time to study, to families, to friends. It is a time-shift into the constant-present, where time is fractionated into nanoseconds, where the past recedes into irrelevancy and a future that is utterly unplannable and unknowable in anything more than the short term.

Part Two

Learning and Earning in the Information Ecology

Chapter Six

The University in Western Society

The university is one of the oldest institutions in Western society. Its roots go back over two thousand years (387 BC) to ancient Greece and the establishment of Plato's Academy, whose fundamental division of primary, secondary and tertiary levels of education remains with us today. Previously, education was purely a private matter, with the state having no formal role. The function of the Academy, however, was to produce morally developed citizens for the benefit of creating an Ideal State—with the state playing a central coordinating role. It also served as a training ground for the elite of Greek society, for the sons of noblemen and statesmen who would themselves go on to become noblemen and statesmen. Education in the Academy was a form of what we would now call "socialization," where young people were instilled with the correct attitudes required of them as members of the elite. At the elementary level, students (boys, girls, common or noble) were taught the morals and values of their society through songs and stories, along with basic mathematics, grammar and music. The secondary level was for the offspring of the elite who could afford the time and would study music, literature, mathematics, astronomy, reason and from around aged 10 to 20 would also participate in military service. Tertiary-level education was for the chosen few, those who would become the "Philosopher Rulers" or "Guardians," whose training would revolve around the study of philosophy, reasoning and logic. The ultimate purpose of the tertiary-level study was the pursuit of Truth.

Plato's Academy flourished until 529 AD when the Christian Emperor Justinian proclaimed it to be a pagan establishment and closed it. The influence of the Christian church hung heavily over the role of the university and the curriculum it taught for the next five hundred years. Indeed, it was not until the beginning of the Renaissance that religious authorities accepted the principle whereby secular learning was seen to have value in its own right apart from theology. Secular learning, however, still had to be licensed by the church, so what developed out of this was the idea that set in place the basic elements of the university that we can still recognize today. This was the notion of the *Studium Generale* divided into four or five faculties: theology, law, medicine, arts or philosophy, and music. Importantly, the university and its faculties were both incorporated and regulated by a self-governing academic body, initiating the long and more than occasionally turbulent principle of *autonomy*. One of the earliest universities to be founded was in Bologna in 1088. The universities of Paris, Oxford and Salerno were founded in c.1150, 1167 and 1173, respectively. By 1400 there were around fifty universities in Europe (Davies, 1997:1245).

Notwithstanding the important principle of autonomy, documents from the Middle Ages show that the four or five faculties around which a university was organized were considered a "unity." This unity was bound by the still-powerful influence of Christian theology. As Nikolaus Lobkowicz (1983:32) notes, the original thinking behind the faculty split and the license given by the church was that:

> theology serves the spiritual goal of mankind, the liberal arts prepare for theological investigation; and law and medicine both serve the social and bodily needs of man.

Institutional autonomy from both church and state, however, was a powerful boost to the culture of freethinking—or at least having the space to think differently from the dominating theological and feudal dogmas. From around the twelfth century onwards, universities began to evolve into intellectual and moral communities that pursued non-theological questions. This led to a return to or discovery of the works of Plato, the neo-Platonists and metaphysics. The pursuit of new forms of knowledge and new ideas based on the Greek classics created its own self-perpetuating momentum. The next four or five hundred years, through the period of the European Renaissance to the dawn of the Age of Enlightenment, saw the development in the universities of a *secular humanism* or *studia humanitas*. Humanism itself was based on Greek classical learning stemming from Plato and was the basis upon which its adherents sought to understand the nature of man in the world and the pursuit of knowledge. As Wallace Ferguson (1962:291) wrote of the rise of humanism and the humanists: "What the humanists sought in the classics was a foundation upon which to build a culture in the broadest sense. What they found in the ancient literature was a liberal education." Importantly, humanism developed most powerfully into the modern period neither as a philosophy or social movement but as an educational curriculum.

From its beginnings, the humanist education program stressed the practical over the philosophical. Its purpose, harking back to Plato's Academy, was to prepare people to lead others and to participate in public life for the common good. Out of educational humanism, then, developed a distinct strain that has been called civic humanism. The civic humanists agreed on the importance of eloquence, grammar, rhetoric and logic, but they stressed political science and political action over everything else.

The Enlightenment saw the rise of civic humanism as one of its guiding principles. For the *philosophes* of the age, it was the key to a more just and fair society. On the local or national level, education would serve to free men and women from the shackles of superstition, religion and backward prejudices, and on the international level *commerce* or *free trade* would lead to a harmony amongst nations and the creation of wealth within and between nations. This was the very stuff of Adam Smith, himself a professor of Logic and chair of Moral Philosophy at Glasgow University, and it was tremendously influential. It is here, at this point in eighteenth-century Britain as the Industrial Revolution was spreading throughout Europe and the world, that the role and purpose of the university in society began to change. It is a moot point, of course, and one I cannot pursue here, whether or not the ancient and venerable "autonomy" of the university, or the university as representing a "community of scholars" who were somewhat detached and aloof from society, ever really existed. What is clear is that from this point onward, in North America, in Britain and elsewhere, the ideas of the civic humanist and the projects of the industrial capitalist dovetailed. Through this intellectual and commercial marriage, the nature of humanism changed. As Walter Rüegg (1983:122) put it:

> The realization of Greek ideas was to be found not in ancient Greek society but in democracy as envisaged by the North Americans. The human ideal was no longer in the Greek youth, but rather in the self-made man of liberal capitalism.

Lobkowicz (1983:31) argues that the university has the tendency to be "quickly overcome by the spirit of the age." And so, during the eighteenth and nineteenth centuries, imbued with a civic humanism that valorized action and utility, universities began to open faculties and offer degrees in subjects such as engineering, chemistry and physics. The

Academies of Science that emerged in most European countries and in the Americas buttressed these, as did various institutes of technology, workingmen's colleges and so on. These faculties and institutions certainly produced knowledge and innovation as well as new and radical perspectives on society, but they were born and maintained in the service of the economy and functioned as the technological underpinning of industrial capitalism. It is here that we find the locus of the seemingly timeless "crisis of the university," one that remains with us today. The persistent argument over the question "What are universities for?" stems from this basic philosophical contradiction. Should they produce wisdom or utility? Can and should they do both? If so, where is the line to be drawn and who is to draw it?

To some extent, the universities in Western society have been able to tread a fine line between these issues, with compromises and trade-offs obscuring but not quite killing off the fundamental question—"What are universities for?" Right up until very recently, guaranteed government funding together with the long-standing privilege of academic tenure, professional prestige and intellectual autonomy kept the humanities and social sciences disciplines from serious open revolt. Let business fund chairs in chemistry or commerce or engineering: "wisdom" and "utility" largely avoided each other, and both more or less got on with practicing their own disciplines, with the "What are universities for?" question only emerging sporadically. Consequently, for much of the twentieth century, Western universities, and English-speaking universities in particular, were able to pursue a fiction that only becomes apparent with hindsight. The question could be subsumed under the compromise. Liberal academics were still funded to teach and research abstruse, economically disinterested subjects and questions. They (mostly) were free to criticize governments and their policies on education as well as the social, economic and political issues of the day. This fiction was especially sustainable during the boom years of Fordism. Indeed, at times during this period, liberal academics and humanities and social science faculties could be said to have been fulfilling their ancient role of criticism, inquiry and the demonstration of a reflexive attitude toward society. For example, the universities in the USA were in intellectual and political turmoil during the latter half of the 1960s, helping in no small part through sit-ins, demonstrations and so on to turn public opinion against the perceived US aggression in Vietnam, Laos and Cambodia. In Europe the French state was severely shaken by an uprising of students, intellectuals and workers in 1968, and left-wing politics gained in popularity and credence through the university-inspired New Left movement in a whole range of developed countries. The "crises of capitalism" of the mid-1970s, however, meant that the day of reckoning for the universities was finally at hand.

We have seen how these crises provoked a radical response from corporate capital and government. The social contract that had endured for much of the post-war period had kept universities, like workers, in a sort of world of unreality. Rising profits and prosperity meant that the role of universities never really came under serious pressure from the business sector or from government. The "fiction" of autonomy and otherworldliness could be maintained as long as profits were being made elsewhere. However, as Jan Currie has noted, the neoliberal form of globalization that emerged out of the crises was underscored by a powerful ideology that has swept almost all before it. Citing the work of Foucault, Currie (1998:11) argues: "Globalization as an ideology has become...*a régime of truth*, which tends to be all-encompassing or *totalizing*." This régime demanded, principally, that the market be the solution for society's ills and that the market and its commodificationary logic be given free rein in every nook and cranny of society. As we shall see, the universities were no longer able to maintain the uneasy fiction of autonomy under neoliberal globalization's particular "régime of truth." Moreover, so powerful has this régime appeared to be that the ancient disputes over the question "What are universities for?" began moving rapidly toward consensus.

The Commercialization of the Higher Education System

Globalization is a dynamical process powered by the twin engines of expanding capitalism and the information technology revolution. A major consequence is that national and transnational capitalism, massively aided and abetted by innovations in ICTs, has been able to penetrate deep into culture and society, into realms hitherto shielded from the market or where previously the market had shown no interest in going. This is a process of colonization and commodification. This globalization/ICT nexus, as Rifkin (2000:97) puts it, produces "commercial networks of every shape and kind [that] weave a web around the totality of human life, reducing every moment of lived experience to a commodified status." Rifkin here puts it rather more strongly than I would be prepared to, but the logic and momentum of globalization and the ICT revolution are certainly moving in this direction. Nevertheless, given the power of the commercial forces and given the immense commercial opportunities that the education sector—and the higher education sector in particular—represented, there was simply no way that it would remain immune from these processes of colonization and commodification.

To understand why the universities were particularly ripe for colonization and commodification we need to view the processes, first from the perspective of the political economy of neoliberalism and its project of restructuring global capitalism upon the basis of free markets and capital's "natural" right to seek profit and investment opportunities without hindrance from governments or allegedly archaic institutions such as labor unions. From the perspective of the "régime of truth" of neoliberalism, markets and market mechanisms are the only fair, productive and efficient way to organize human relationships—all of them. In this respect, universities were seen to be as much in need of economic restructuring as were the nationalized and subsidized industries that had allegedly grown fat and sluggish during the post-war boom period. Of course the universities could not simply be sold off or desubsidized as many industries were during the phase of economic restructuring. They first had to become amenable to the "laws" of market forces. What was necessary to begin with was a change of attitude at the senior levels of the university, in government and in the general population. The details of how universities were placed on an entrepreneurial, market-oriented basis differ from country to country, but only in detail. The *aim* in all Anglo-American countries was the same. There is no need to go over the policy developments here as they are fully analyzed in case studies of two trailblazing countries that began to commercialize their universities in the early to mid-1980s: in Australia in Marginson and Considine's *The Enterprise University* (2000), and in the USA in Sheila Slaughter and Larry Leslie's *Academic Capitalism* (1997). The common institutional effect in the universities was that they came to see themselves (or were forced to see themselves) as businesses that had to compete in the marketplace like any other. Government subsidies were still keeping them afloat, but these were constantly being reduced and were, as in Australia, Britain, Canada and New Zealand, now subject to market-based "performance indicators" of "efficiency." Since at least the mid-1980s the university in the Anglo-American sphere has been in a constant flux of reorganization, restructuring, amalgamation and expansion. Indeed, as universities now operate largely along business lines, this is how they *must* operate as long as they continue to function as commercialized entities.

Macro-level policy developments have effected deep institutional change at the individual university level. In short, the university is now run in a radically different way. It essentially apes the transnational corporation. Power has concentrated at the very top of the organization, to the office (and more usually the person) of the vice-chancellor (VC). Marginson and Considine (2000:9–10) detect five trends in governance that are common across all universities in Australia. These largely follow global trends. First, the new VC is part of a generic executive breed. They tend to think and operate using common language and assumptions derived from 1980s and 1990s US and British management theory. The dogmas of "enterprise culture," "efficiencies," "world's best practice," "value-adding," "branding" and so on have been internalized as the principal reality and act as strategic cues for their actions. They have learned how to "talk the talk and walk the walk." This has produced its moments of black humor, such as when Princeton was caught dealing in a little "industrial espionage"—it hacked into the computers of business rival Yale to view its files for online student enrollments (*New York Times*, 14 August 2002).

Second, structural changes have remade or replaced collegial or democratic forms of governance. VCs and the senior executive group work at arm's length from the faculties, concentrating central executive power, dictating the formal agenda and pursuing outside business ventures, often in "sensitive and lucrative areas," without discussion or consultation or collegial debate. Third, as in the corporation or company in the age of globalization, "flexibility" in staff and in resources is seen as the *sine qua non* for "world's best practice." In terms of staffing, fixed-term contracts have become the norm, as sessional short-term contracts (yearly, semester to semester or, in this author's experience, even week to week) get much of the teaching done. Use of resources and staff "are no longer governed by legislation [but] by formulae, incentives, targets and plans." Fourth, at the faculty level, academic disciplines with their remaining vestige of collegiality and professionalism are "often seen as a nuisance by executive managers and outside policy-makers." Top-down concentration of power has left the faculties susceptible to constant restructurings and the break-up of disciplinary identities and structures. Moreover, divide-and-rule is the outcome if not the intent behind funding and performance systems that drive academic work via "a common disciplinary model, flattening out the distinctions between different kinds of knowledges, while enabling the university centre to reach directly into the work program and resource base." And fifth, concentration of executive power and "devolution" of responsibility become part of the same process of centralization. Lower levels of management, deans, faculty heads and department heads have been given "budgetary autonomy" within the framework of strategic "formulae, incentives, targets and plans." These lower levels are also partly responsible for generating outside funding. A major effect of this devolution is that the pressure for the overall strategic plan to work falls upon the line managers—as does the blame for any failure.

This new reality stems from the universities' relationship with the state, and in all the Anglo-American countries this has followed the pattern of cutting subsidies, forcing them onto the market and to private sources of funding to make up for the shortfall in what is a massively expanded higher education sector. This basic change has necessitated the formation of business-oriented and anti-democratic organizational structures within the universities that have kept them in a constant state of flux as they attempt to keep pace with changing market conditions. It has also brought about an intrinsic reordering of what universities do. As Claire Polster and Janice Newman (1998:174–5) have argued, the introduction of "performance indicators" tied to government subsidies and private sources of funding has radically changed the nature of teaching and research. Their effect has been to "manage and control the academic activities that flow within and through institutions of higher education. As such, they are conceptual devices that link academic judgment to budgetary and policy-sensitive constructs." The judgment of teaching quality, for example, a process formerly developed through discussion and debate within collegial bodies, is now widely determined through "mechanically produced and standardized 'facts' such as class size [and] student output—to assess the cost-effectiveness of a given institution's deployment of its teaching resources." What this means is that teaching quality is being assessed through accounting procedures. After the colossal corporate accounting malfeasances in Enron, WorldCom, AOL Time Warner and who knows how many others, this does not seem a sound basis upon which to organize the core of what universities are about. In university research, the effect has been that inquiry for its own sake is no longer seen as a pre-eminent function. There has been a profound shift toward conducting the kind of research that government and business deem economically useful—as their accountants would naturally suggest. Accordingly, academic research has become what Slaughter and Leslie call "resource dependent." This, they argue, has been fundamental in the shift toward "academic capitalism" where universities now have to compete and "engage in market and marketlike behavior" in order to attract resources (1997:114). In a very short space of time, the institutions of teaching and learning have become businesses, and, therefore, "competition...becomes the basis for the logic of discovery" (Delanty, 2001: 108). Moreover, in line with the general patterns of globalization, these have become globalizing businesses with many VCs eagerly adopting the role of CEOs. They now spend much time empire-building, merging with other universities, expanding overseas, partnership-building with industry and, most of all, chasing the "gold mine" that is the education of fee-paying international students, worth an estimated $12.3 billion annually in the USA alone—part of the rationale behind Princeton's hacking activities, presumably. ICTs are central to the whole process. A European Commission (2000:9) research paper entitled *The Globalisation of Education and Training* puts the case in terms blunt enough for the entrepreneurially minded VC to get excited about:

> As world-wide communication becomes easier with the help of ICTs, the idea of conquering the world market of educational products and services is increasingly attracting business-minded established institutions and profit-based new providers in a race in which most traditional higher education institutions have as yet been left far behind. The emerging market of educational products and service is generally regarded as a gold mine. (italics in original)

The commercialization of the university is primarily an economic and political process of transformation that has little if anything to do with education, knowledge production and the well being of either staff or students. What is more, these changes are all being refracted through the prism of neoliberal ideology. The economic dimension was justified in terms of the new realities of the post-Fordist age: the university simply had to be more responsive to the needs of the economy. In the new hyper-competitive climate that is neoliberal globalization, the university, like every other institution and individual, had to become "flexible" and market responsive. As government subventions decreased, universities looked, necessarily, to the market and the private sector for alternative sources of funding. Over the 1980s and 1990s the idea of the university as serving the broad-based public good began to be defined in narrow economic terms: the "bottom line" (Marginson and Considine, 2000:28).

This narrowing of the key concerns of the university has been reflected in what students go to university to learn about. There has been a tremendous growth in business and vocationally oriented courses and degrees, leading to charges that the university has become a training ground for industry and little else (Hoad, 2000). Postgraduate courses offering Masters in Business Administration (MBA) awards have become lucrative profit centers for many universities. An average US university can charge a student around $US8,000 per term, whereas in academic year 2001 Harvard Business School invoiced its yearly intake of 880 students for $US28,500 each, guaranteeing for itself $US25 million from a small part of what it does and whom it charges. In undergraduate courses, the number of students enrolled in technical and business subjects such as hospitality, accounting, tourism, management, marketing, computer science and software engineering far outstrip the number enrolled in "traditional" humanities and social science courses such as politics, philosophy and literature.

The over-riding concern with the "bottom line" has meant that in most Anglo-American universities the cost of education has begun to shift to the individual student, and so fees are now being levied for the privilege of studying these subjects. Economic logic being what it is, most students are unwilling to pay for a university education that holds out little prospect of getting them a job at the end of it, so the rush to vocational subjects increases as the attractiveness of vocationally disinterested courses dwindles correspondingly. The "gold mine" of the fee-paying international students that most universities pursue energetically is, needless to say, almost wholly business oriented. An important consequence of this is that the massive preponderance toward business and vocational education has developed its own ideological dialectic. The millions of students from across the world graduating from these courses, especially after paying for them, will through their training internalize the cash nexus as the only reality and the market as the only rational form of economic and social organization. As this process of ideological re-education continues, diversity in ways of seeing and of being will become relentlessly narrowed and one dimensional, and the pressure to conform to its logic increasingly difficult to resist.

The commercialization of the university has certainly been a revolutionary process but not a "cataclysmic" one as Peppers and Rogers might have termed it. It has been a quiet revolution, though with its share of struggles and micro-dramas. The transformation to commercialization has foreshortened many academic careers. Much "dead-wood" was cut away in order to let the green shoots of entrepreneurialism spring forth. However, other careers have been opened up, and many perspicacious and flexible academics and administrators saw the writing on the wall and proceeded down the grim existential path of intellectual and professional reconstruction in order to adjust to the new realities.

The university is an institution that has changed only glacially over the centuries. In the last twenty years, however, it has metamorphosed rapidly into a completely different institution—if such a perpetually mobile business-oriented entity may still be called an "institution." So radically has the university changed that the typical academic, administrator or student from the 1960s and 1970s would barely recognize it today. It might seem to them to be more akin to a marketing company or advertising agency, so concerned is it with profit, products, clients, market share, branding and image. Another aspect of university life would seem even more alien—the automation and computerization of almost every aspect of its functioning.

The Informationization of the Universities

In his important and passionately written *Digital Capitalism: Networking the Global Market System* (1999), media theorist Dan Schiller looked at the forces that led to the "networking of the higher-learning industry" in the USA during the mid- to late- 1990s. The following quotation is reproduced at some length as it encapsulates, succinctly, some of the processes and their outcomes I have described and will describe further in this section. Getting seriously into the stride of informationization process by the 1990s, Schiller (1999:144) argues that:

> The Internet had been overlaid on a domain—education—that was itself already awash in change. Indeed, it was becoming apparent that the entire established system of skill formation and knowledge creation was heading for a makeover. Where once there had existed nonprofit institutions, increasingly, there were now commercial vendors. Where once there had existed relatively autonomous instructional and learning processes, increasingly, there were now attempts to cater more directly to labor markets. The system of education provision was being reoriented toward familiar corporate practices that were foreign to the bulk of earlier educational endeavor: growing utilization of casualised labor, productivity enhancement measures and product development based on profit and loss potentials. A concurrent and related reform, toward school-to-work programs,

lifelong learning, and "new partnerships," symptomatized an intensifying vocationalization of the education process.

It is important to be clear at the outset where this impetus for the informationization of the university and the corporatization of knowledge production came from. To be sure, there was never any groundswell of demand from faculties, from heads of department, from students and from university administrators for "more computers," "email," "networks" or "online learning." At the beginning of the 1980s, even senior administrators and VCs were not to be heard banging the drum for the automation and computerization of the education system. At this point in time, computers were still considered by most people to be fairly exotic. If people gave the matter any thought at all, computers would be thought to have something to do with NASA and other high-technology pursuits, not the study of politics, education, economics or even business and marketing. This is not to say the universities were oblivious to computers and computerization. In many cases, they were at the cutting edge of research and development in this field, but the computer's application was perceived as limited and specific, for this or that task or problem not as the solution to all the world's concerns and difficulties. There were distant rumblings outside the university, however, and these were getting nearer all the time. A revolution was underway, with its sights trained on the academy, and soon the twin forces of neoliberalism and the information technology revolution were to kick the doors down.

As Theodore Roszak tells it, the high-pressure hype of the "data merchants" from computer multinationals, backed by futurologists and industry consultants, established the initial bridgehead into the universities. There was no empirically demonstrable need (then or now) for the blanket computerization and automation of schools, colleges and universities. It was, argues Roszak, fundamentally about "selling" (1986:31). Public awareness of the new Information Age was just beginning during the early 1980s, an awareness that was buttressed and hyped immeasurably by the influence of best-selling books such as Alvin Toffler's *The Third Wave* (1980) and John Naisbitt's *Megatrends* (1982). Information technologies and free markets were the new waves of the future, we were told. The message from these books and a hundred other derivatives was woe betide any company, any institution, any government or any individual who fails to recognize this new reality. Computers (in tandem with the market, of course), the data merchants told us, could do anything, and represented the future of everything. In practice this meant that computers came to be seen, as Roszak memorably puts it, as "a solution in search of problems" (1986:51). In the restructuring of Western capitalism that took place throughout the 1980s and 1990s, governments, business leaders and consultants identified a problem—usually a perceived economic "inefficiency"—and applied an ICT "solution" to it. In the tumult of market-derived paranoia, schools, colleges and universities were also seen as in need of urgent attention. Futurologists, business gurus and business executives were continually telling the Anglo-American governments—and anyone else who would listen—that the Japanese and the rising Asian "dragon economies" were soon set to eclipse all other countries, breaking a two-hundred-year-old technological, industrial and cultural domination by the West. Such fears re-energized the long-standing "what to do about the crisis in the schools" debate, and so the information technology "solution" was eagerly offered by the computer industry and embraced with alacrity by schools, universities and government.

Computer companies argued that computerization and automation would have twofold benefit for educational establishments. On the one hand, students would become "computer literate" and develop skills that would equip them for competition in the new information economy. On the other hand, ICTs were more "efficient" than teachers and would revolutionize the ways in which teaching and learning was done. Much of this sales pitch was taken at face value by local educational authorities and by apprehensive state and national governments in the Anglo-American countries that were continually looking over their shoulders for signs of the "Asian threat." And so by the mid-1980s the mass computerization of the education sector was underway and has been continuing apace ever since, as spending on books has declined or stagnated.

How effective has this massive investment been? What, exactly, does "computer literacy" mean? Does it mean the ability to use a computer? If so, in what ways, other than knowing how to use a computer, would this help a student? What about the computer's much-lauded use as a "learning tool"? More important, what type of learning "works" and which subjects are best suited to the rapidly accelerating information ecology being created in the universities? A report by the German government's President's Information Technology Advisory Committee (PITAC, 2000) displayed an admirable caution when considering this question:

> We know too little about the best ways to use computing and communications technology for effective teaching and learning, in particular, how to effectively use multimedia capabilities to create a richer and more appealing learning ex-

perience. *We need to better understand what aspects of learning can be effectively facilitated by technology and which aspects require traditional classroom interactions with the accompanying social and interactive contexts.* We also need to determine the best ways to teach our citizens the powers and limitations of the new technologies and how to use these technologies effectively in their personal and professional lives. (italics added)

No such reflexive attitude prevails in the USA, in Britain, in Australia or any of the other Anglo-Saxon neoliberal countries. When asked about what discussions had taken place regarding the possible downside of computerized education by members of the US Presidential Technology Task-Force, the response was that "there weren't any" (Oppenheimer, 1997:10). Hardly any basic research has been done in this area, either during the 1980s or today. Such research as has been done, if found to be equivocal or negative regarding the interaction of computers and human learning, runs the risk of being disowned by its sponsor. For example, researchers from Boston College, commissioned by the Massachusetts Department of Education, found that students who have learned to write using a pencil and paper perform better than those who have learned on a computer. The Department, according to Michael Russell, the lead researcher, "resisted" release of the study, eventually requesting that its name be taken off the published findings altogether (Carlson, 2000:A40). Similarly, the benefits of learning to read and develop comprehension skills via computer programs are difficult to pinpoint. Todd Oppenheimer (1997:9) cites one study done in the USA:

> One small but carefully controlled study went so far as to claim that Rabbit Reader, a reading program now used in more than 100,000 schools, caused students to suffer a 50 percent drop in creativity.... After forty-nine students [had] used the program for seven months, they were no longer able to answer open-ended questions and showed a markedly diminished ability to brainstorm with fluency and originality.

Data merchants also sold computerization as a "more efficient" way to teach students. Lectures could be stored on laser disc, endlessly replayed to an infinite number of students, anywhere, anytime—in the comfort of their own home or in a lecture theatre. Considering this issue, Roszak (1986:54) mused:

> Why should we want to invent such a machine in the first place? There was never any difficulty in answering that question when the machine was intended

to take over work that was dirty, dangerous or backbreaking. Teaching is hardly any of these....

The answer is, of course, market forces and market pressures, not any verifiable, empirical and unambiguous evidence about their utility as learning and teaching tools. Possibly never before in history has so much money, so much investment, and so much radical social, economic and cultural reorganization gone into something so poorly understood. Aside from their undoubted number crunching and batch-processing capabilities, it is by no means clear what indisputable benefits computerization has brought to the university. Yet the process of informationization, driven by market forces and by universities attempting to swim in their fast-flowing currents, simply continues and increases in scale, complexity and density. What is clear is that computers are very versatile and their logic can be applied to just about anything—which is precisely what the industry has done in order to sell as many computers (and the ever-expanding panoply of consumables, add-ons and upgrades) as possible.

And the computer market in the education sector is a very big market indeed. In the USA some $US1.3 billion was being spent on computers in 1984 (Roszak,1986:57). By 2010 the global ICT market value in the education sector is predicted to be $US4.5 *trillion* (Stewart, 2001:2). The university sector across the world now represents hundreds of millions of personal computers in labs, offices and administration areas. These are connected and connectable to immensely complex networks within and between universities and are constantly being upgraded, replaced and added to in the ongoing informationization process. These, in turn, connect to the Internet, opening up millions of users to each other and to hundreds of millions of pages of data, writing, research, argument and entertainment. So densely informationized are most Anglo-American universities today that it is not possible to go to one without what only ten years ago would count as fairly sophisticated computer and multimedia skills.

Competition makes the thickets of interconnectivity grow denser still. Following the lead of the multi-billion-dollar IT education plan in the USA, the European Commission in 2001 approved what it terms the "e-learning Action Plan," a $US13.3 billion project to boost ICT-related education. The informationization of the universities necessarily makes new demands and develops new learning practices in students. Many universities advise prospective students that a laptop or desktop computer is now an essential tool and offer schemes whereby they can purchase one at discounted rates. Indeed, in Massachusetts state schools and colleges, ownership of a laptop is now mandatory for all new students. Increasingly universities invite students to enroll online, change courses online and pay fees online. More and more subjects are being delivered (in part or in whole) online; from an undergraduate degree through to masters and Ph.D. levels, it is possible now to undertake the entire course, and hence the entire university experience, in front of a PC in your bedroom or wherever. Students are encouraged to communicate with lecturers and fellow students online through email, bulletin boards and text-messaging; course materials are increasingly available only online, with students required to access and download them each week; end-of-semester results are now widely available online; some universities in Australia are sending students their results through cell-phone text-messaging, and in Holland, at the University of Enschede, 10,000 students were given free WAP-enabled cell-phones by the Ericsson Company and Lucent Technologies to connect them even more intimately to the real-time learning network (Dorsey, 2001).

In keeping with their status as near-businesses, another logical step for universities in the age of globalization is geographic expansion. Just as the Ford Motor Company devised its global car strategy around ICTs, so, too, do universities place information technologies at the center of their strategic plans. The "internationalization" of the university is the current trend stemming from its mimicking of the business corporation, and any university wishing to be taken seriously is involved in it in some way. In the attempt to capture the international market at source, many are now actively physically expanding overseas. For example, Australian universities have established campuses in Asian countries such as Indonesia, Malaysia and Thailand or have collaborative arrangements with institutions in Hong Kong, China and South Korea. Many other Anglo-American universities are involved in these expansions, such as the University of Nottingham in Malaysia, and the collaborative project between the US Wharton Business School and Northwestern University, which established the Indian School of Business in Hyderabad, to name but a few (Thornton, 1998). The major selling point is being able to study for a "prestigious" Western degree without having to leave home. A charitable view of this globalizing process would be that ICTs have enabled universities to fulfill some kind of altruistic educative mission for those students from less wealthy countries. As Elaine Martin (1999:11–12) explains it:

> Internationalization of higher education is not just about Western nations providing education for less-developed nations and, consequently, increasing revenue. Underpinning the notion of internationalization is the idea of developing a shared understanding of cultures and politics and markets.

Increased revenue is thus only a consequence, almost an afterthought. This interpretation, however, runs counter to the logic of why businesses are in the business of business in the first place: think back to Adam Smith's dictum that the baker does not bake bread for us to eat but for himself to make money from. This view also sits uncomfortably with the fierce competition between universities for access to this particular "gold mine." Moreover, it studiously avoids issues of cultural, economic and political imperialism—where are the Asian university campuses in Australia or Britain or the USA to help develop a shared understanding of each other? A somewhat less generous (and more accurate) view of this process is that they are first and foremost revenue-generating ventures, with a "consequence" being that new generations of Asian students are being inculcated with Western business values of competition, individualism, the sanctity of market forces and (implicitly) the superiority of the Western way of life.

These onshore and offshore digital worlds run parallel to the new cyberworld of the "virtual university." As most universities now consider themselves in the "business of education," the virtual university is simply the next logical step, mimicking as it does the corporate-sector fascination with the possibilities of the "virtual corporation." This particular idea stems from 1990s management theory and has been instrumental in the transformation of many multinational corporations seeking to make themselves "virtual" and "weightless" (Davidow and Malone, 1992, Rifkin, 2000). The theory states that to be successful in the New Economy like Microsoft or Nike, corporations have to concentrate on intangibles such as "ideas" and "brand," instead of old economy assets such as buildings, plant and machinery. Unsurprisingly, market-savvy executives in major universities thought they were perfectly placed to exploit such a way of doing business. As the name suggests, the virtual university does not exist in the traditional sense. It is purely digital, existing nowhere but potentially everywhere, it is both virtual and weightless. Distilled down to its essence, for the "bottom line" university executive, the virtual university makes perfect sense. What does the university provide?—knowledge and ideas. These are not bulky or tactile or consisting of mass and so are eminently digitizable and marketable in the Knowledge Economy. Importantly, these are not costly overheads like real estate, buildings and staff, and expansion into the virtual realm will thus represent immense savings. Moreover, in the fast-paced world of competition, where time is of the essence, the virtual university student can access learning materials when he or she is able, anytime, anywhere; no need to attend classes or visit libraries, have meetings with lecturers or deal with boring administrative matters—it can all be done virtually in your spare time. Elite universities have another valuable intangible they are able to bring to the mix—brand recognition. Universities such as Harvard, Yale, Oxford and Cambridge all have virtual university or extensive online projects either in the pipeline or up and running, using their brand names as selling points. As with the international education market, competition in this marketplace is fierce. This has had some interesting consequences. In 2001 the Massachusetts Institute of Technology (MIT) decided to allow free and open access to all its online material. Students can freely download and study all its material—but to obtain an MIT degree, he or she, of course, must pay. MIT is banking not so much on its belief in the quality of its material but, possibly more importantly, on the "brand" that is "MIT" to entice prospective students to sign up for a course.

Brand recognition is a major asset for corporations such as Coca-Cola, McDonald's, Sony and so on, and the entrepreneurial (and branded) universities have been able to market themselves globally on the basis of brand value. If less globally recognizable universities are to compete for market share, *brand building* becomes part of the virtual university strategy. Here collaborations and partnerships have developed in the attempt to construct a global image of excellence. An example of this approach can be seen in the Universitas 21 consortium, a collaborative project consisting of eighteen universities from Europe, North America, East Asia and Australasia. The project represents virtuality in its purest form thus far within teaching and learning. Physical universities such as McGill, New York, Glasgow, Freiburg, Melbourne, Hong Kong, Auckland and others are joining together online and lending their individual scholarly legitimacy (such as it is) to create an electronic brand, an entity that exists nowhere and has been manufactured with "brand-building" as its primary objective. As Universitas 21's website (2001) unabashedly proclaims:

> The Company's *core business is provision of a pre-eminent brand* for educational services supported by a strong quality assurance framework. It offers experience and expertise across a range of vital educational functions, a proven quality assurance capability and high brand value. (italics added)

Branding, as is evident from this passage, is central to the project, and, not un-coincidentally, branding has been one of the management theory crazes of the 1990s and beyond, underpinning the New Economy and the "virtual" and "weightless" corporation. Branding is a business tool around which the Universitas 21 virtual university is constructed. And New Economy gurus see the brand as an almost spiritual relationship between producer and consumer—consummated in the logic of the market. An angry and ironic critic of this process is Thomas Frank (2001:254):

> The brand, according to the new cognoscenti, is a relationship, a thing of nuance and complexity, of irony and evasion. The brand is a conversation, an ongoing dialogue between companies and the people. The brand is a democratic thing, an edifice that the people had helped to build themselves simply by participating in the market. The brand, in short, is us.

What does brand-building mean in relation to the virtual university and the market? It means that for the new entrepreneurially minded universities, the brand, the business, the market, the consumer and the cognitive processes of learning are all able to seamlessly interact in the New Economy, with each dimension simply augmenting and complementing one another. What is more, the shift to virtuality in the business of learning is inexorable, according to the "new breed" VCs. According to Alan Gilbert, University of Melbourne VC and Universitas 21 booster, "we face a future in which it may be increasingly difficult to argue that the benefits of face-to-face scholarly interaction are not offset by the relatively lower costs of 'virtual' interaction" (Gilbert, 2000). Gilbert, like many other "new breed" VCs, is explicit about where the motivation for online delivery comes from: the marketplace. As a market-driven and profit-seeking project, it is therefore natural enough that the initial offerings of the Universitas 21 consortium will be in the business area: the sale of MBAs. And in the education marketplace, online delivery of MBAs is lucrative. Students, for reasons I shall discuss in more detail later on, seemingly *prefer* online courses in areas such as the MBAs, notwithstanding the fact that, for example, the University of Florida's Warrington College of Business is able to charge *three times* as much as the traditional face-to-face MBA program (Baum, 2001).

To end this section, having charted the contours of the commercialization and the informationization of the universities, we may ponder the question: what does it matter if universities become entrepreneurial and use ICTs to their fullest possible potential? To many it may indeed be a positive development. Universities at last have been made to pay their way, to come down from the ivory tower and become democratic, responsive and inclusive. After all, the higher education sector has massively expanded, right across the developed and developing worlds, bringing education, skills and opportunities to millions who previously would have gone without. This is indisputable. However, there are significant and serious downsides to this process. The rapid evolution of universities into commercial entities has changed their background logic. This is a tradition that valorized diversity, independence and the pursuit of disinterested inquiry. This essentially was knowledge for its own sake, something that makes the totality of human experience deeper, richer and more reflexive of itself—*as well as* able to make contributions to economic development.

Universities that are dominated by the commercial imperative, entities that are "resource dependent," are necessarily narrower in their outlook, restricted in their room for maneuver and experimentation, and one dimensional in their worldview. Here the world is represented as markets, market share, profit and value—and opportunities to exploit these. The product is knowledge, but this knowledge, in the main, needs to be market responsive, that is, vocational, and research, the creation of new knowledge, boils down to what can attract funding. Mohammad Hamza and Bassem Alhalabi (1999) argue that emergent ICTs in the university are marked by their "heterogeneity" and that online delivery in the commercial context amounts to "poorly structured academic programs…that entice students to pursue 'fast-track diplomas'."

It should be noted that the curriculum developed for Universitas 21 and other "virtual universities" and much general online delivery comes from non-academic education software providers such as Thompson Learning. Accordingly, teachers (or "customers" as the Thompson Learning website calls them) have no input into what they will be responsible for delivering online—they simply operate the software and transmit its contents to students. Citing research on these marketing strategies, Hamza and Alhalabi (1999:2) go on to describe these online degrees and diplomas as:

> little more than moneymaking plots managed by capitalistic-minded individuals who held verily the slightest regard for academic values. Their academic services lack academic authenticity and education quality.

The argument that this process is some sort of "plot" is not tenable. A plot would take the collusion of all the universities and governments to hoodwink gullible students. Lack of quality, or any degradation in academic standards, is the simply the consequence of putting education in the marketplace, a selling of degrees as commodities (Noble,1998). *Authenticity* is certainly an issue as students experience education and learning both virtually and rapidly. The great danger of the virtual experience is that as new cohorts of students enroll and emerge out the other side of a degree course in the new commercialized environment, their thinking will have become one dimensional because this is how they have been trained—authenticity has been replaced by expediency and convenience. The inculcation of skills and habits of critical and reflexive thought are degraded, and this spills over into civil society and our knowledge of ourselves as part of a community, as part of a region and as part of an interdependent and interconnected world. Difficult questions are raised by the rapid evolution of this system and the logic that sustains it. In a digitally accelerated world of earning and of learning, a world where a sense of history and a projected future seem less and less relevant, will we still be able to think hard and critically about the world and our place in it? Will we still be able to make the connections between such things as the cause and effects of injustice, the cause and effects of the concentration of economic power, of the widening disparity between rich and poor, of environmental degradation, or not being able to recognize our next door neighbor any more?

The coming to pass of an end to the inculcation of critical thought and reflexivity in successive generations of students may be in sight. The demise of the university as "a foundation upon which to build a culture in the broadest sense...a liberal education" that can function as an inestimable social institution is certainly upon us. The neoliberal "régime of truth" is producing its own authenticity—an ideologically contrived authenticity that states that "truth" is the "bottom line." The traditional and venerable stature of the university as a bastion of secularism is thus being undermined by a belief in the religion of the market, of capitalism and information technologies as the solution for society's ills, its injustices and its inefficiencies.

The immense social, cultural and political problems that these trends foreshadow do not simply emanate from the commercialized and informationized university where a one dimensional and vocational education is increasingly all that is on offer. They are society-wide and stem from the wider changes that have emanated from neoliberal globalization and the ICT revolution. As Dan Schiller (1999:205) concludes:

> The common link between [the emerging network society] is a secular buildup of transnational corporate power to define and shape social institutions. A couple of thousand giant companies—as employers of workers laboring on networked production chains, as advertisers and, increasingly, as educators—today preside, not only over the economy but also over a larger web of institutions involved in social reproduction: business, of course, but also formal education, politics and culture.

Accordingly, the student who enters university today is a different human being—psychologically, motivationally, economically, culturally and ideologically—from the "typical" student of twenty or so years ago. And so the raw material that the commercial university now uses to shape the future knowledge workers, specialists, capitalists, and politicians will be the subject of the next chapter.

Chapter Seven

The New Universities and the Student-Worker of the Twenty-First Century

> As soon as you're born they
> make you feel small, by
> giving you no time instead of it all.
> John Lennon, *Working Class Hero*

The Mechanics of Change

Restructuring in the Western economies has been a root-and-branch affair that left no industry unaffected; so much so that the core elements of what sustained the post-war economy began to change. Heavy industries either evaporated into bankruptcy, or they relocated, or became "lean and mean" shadows of what they once were. What began to emerge from the ashes of restructuring and deindustrialization in the late 1970s and early 1980s was a new, post-industrial, post-Fordist economy. This New Economy was basing itself on the high-technology industries such as computers, semiconductors, sophisticated capital-intensive manufacturing and media. These industries were seen to be the industries of the future, and in this early phase of globalization and the information technology revolution, Western corporations were world leaders. To enable them to stay at the front of global competition they argued that they needed to be able to manage their affairs by themselves, without interference from organized labor or bureaucratic government—economic and political processes which we discussed earlier. As far as industry was concerned, though, government still had a very important role to play in the development of the economy. This role was not simply to get rid of red tape, bureaucratic restrictions and "uncompetitive" tax policies; government also had a responsibility to create the appropriate labor market conditions in which the New Economy could thrive and prosper. This meant that skill levels of the working population had to be increased. Totally new skills were needed, sometimes on a fairly large scale, in areas such as in computer science, software engineering, programming and systems analysis. Existing skills in the workforce such as business administration, design, marketing, accounting and so on were also needed, albeit in revamped and reenergized style, as befitting the conditions and characteristics of the New Economy. Note that all these requisite skills, the tools for the New Economy, are *knowledge*-based. Society was moving from an industrial economy based on the production of "things" to one based on the production, commodification and distribution of knowledge (Reich, 1991). In this society most workers will be, in one form or another, knowledge workers, or as Reich terms them "symbolic analysts" in an information economy.

In the period from the end of the Second World War until around the 1970s, the role of the university in most developed societies was, as argued earlier, rather ambiguous. On one hand, the fiction of autonomy could be maintained in the period of economic plenty, and the liberal arts tradition was able to contribute to the "national dialogue" and be a central part of the building of a civil society, where the aims and the values of the "good society" could be developed, refined and articulated. Indeed, the university as the seat of national consciousness moved from the abstract realm of theory and debate, to protest and public disorder in campuses around the developed world in response to the Vietnam War during the 1960s and 1970s. On the other hand, the university also had its strategic uses—especially in the USA and Britain, where the Cold War kept science labs, and graduates from these labs, busy in well-funded work on space exploration, rocketry, atomic weapons development, sophisticated computers and so on (Lowen, 1997). The university only had tangential links to direct economic utility, and this was seen mainly as a *part* (if for some a rather disagreeable part) of what universities were supposed to be for. However, as the crisis of Fordism evolved into economic-political "solutions" of neoliberal ideology of the 1980s, higher education establishments, as traditional centers of knowledge production, came rapidly to be viewed as a vital asset by both government and industry in the shift to a knowledge-based economy. Commenting on the US experience, Richard Florida (1999) observed that:

> During the 1980s, the university was posed as an underutilized weapon in the battle for industrial competitiveness and regional economic growth. [It was] argued that the university had a civic duty to ally itself closely with industry to

improve productivity. At university after university, new research centers were designed to attract corporate funding, and technology transfer offices were started to commercialize academic breakthroughs.

In the race for competitiveness, this trend was widespread, indeed universal, throughout the developed world. A major policy platform in the reorientation of the universities was the OECD and its various education policy fora. The OECD held its first ministers' conference in 1978, prefiguratively entitled, *Future Educational Policies in the Changing Social and Economic Context*. In his book *The OECD Perspective* George Papadopolous (1994) has written that this conference was notable "for the force and urgency with which educational change was advocated to respond to the new economic imperative, marked by growing country interdependence and competition in the global economy" (cited in Lingard and Rizvi, 1998:262). As Lingard and Rizvi argue, throughout the 1980s with neoliberalism in the ascendancy everywhere, the "relationship between economic and social purposes of education [became] explicitly interlinked" (1998:263). As Lingard and Rizvi show, a whole series of OECD education policy conferences throughout the 1980s and beyond were able to create the context of the education-economy link in such a way as to give the impression that there were no alternatives. Quoting Michael Apple (1994) the authors wrote that "significance of these [OECD conferences] lay not so much in the details of the discussions as in the overall orientation of its analysis and its linguistic strategies in creating a rhetoric of justification for a tighter connection between educational systems and the world economy" (1998:263). The education ministers of national governments were thus able to present legislation on the overhaul of the universities to reorient them to the needs of industry as a *fait accompli*, one that applied to every economy (if we don't act now we'll be left behind) and, moreover, had the imprimatur of the OECD.

A corollary to this process, of course, was the massive expansion of the higher education system. Neoliberal globalization has expanded capitalism tremendously, and the pressure was for the higher education system to grow to meet the demands for technical and business skills. And grow it did. An example of the nature, size and pace of this growth can be seen in the case of the Australian education system. Indeed, the Australian system may be said to be prototypical, as it has blazed a trail in attempting to make higher education "pay its way" through the market and make students pay their way through the introduction of a fees system, innovations that have been admired and/or taken up in many other countries. The process has been dubbed the "Dawkins Revolution," after the Labor government Minister of Employment, Education and Training, John Dawkins, who introduced the initial legislation. The ministry responsible for implementing these changes was an new one and one whose very name clearly signaled the future orientation of the higher education system. In July 1988 a White Paper entitled *Higher Education: A Policy Statement*, was released. Once in place, as Marginson and Considine (2000:29) note, the new legislation:

> was also the genesis of a new logic of sector-wide governance which was to deeply inscribe itself upon the psyche of each contemporary bureaucrat and university leader drawn by its siren call.

It was also symptomatic of neoliberalism's ideological narrowing of everything down to the "bottom line." Universities were henceforth to be viewed as industries and run upon market-based lines. They would be rationalized, amalgamated and reorganized from top to bottom to make them "market responsive" and prepared to meet the needs of industry. "Performance-based" payments to universities meant that there was little room for senior management to move. Dissention or obstruction could, in theory, mean the withholding of government subsidy. And, in any case, previous changes to the higher education system had changed the structures of university senior management, and many of the new-breed vice-chancellors were already in place and embraced the revolution with alacrity. The effect was a trickle-down management by fiat, which "university leaders soon began to practice upon their own institution with all the vigour which Dawkins had brought to his attack upon them" (Marginson and Considine, 2000:35). There were carrots as well as sticks. Student numbers were to be massively expanded. Higher education enrollments in Australia increased in the decade 1987–1997 from 390,000 to 670,000. Colleges and Institutes of Technology were able to call themselves "universities," and their senior staff had bestowed upon them the opportunity to have themselves addressed as "professor." The expansion and total reorganization of the Australian higher education system to meet industry needs was in effect government using the universities as tools of micro-economic reform in the evolving post-Fordist world. Importantly, the new market-friendly higher education system was able to help develop labor market and employment "outcomes." To quote Marginson and Considine (2000:29–30) once more:

> [The Labor government] created 16 new universities through upgrading and merger in four years, forced high school retention rates up in most states and convinced a new generation of high school adolescents that university was the logical choice for a post-school commitment. Certainly this helped Labor manage a youth unemployment problem which threatened their social democratic credentials.... New entrants were simply allowed to flow into existing courses and institutions with little attention to career options or ways to mark out a new pathway for less academically formed students. A host of "generic brand" programs in business studies and professional writing began to proliferate in suburban and regional campuses. "Full education" emerged as a substitute for "full employment" and a temporary alternative to life on the dole.

And indeed new waves of student intake seem to bear the same marks of the new logic that has inscribed the psyches of university administrators and senior executives. This particular neoliberal "trickle-down effect" has fundamentally influenced students' reasons for going to university in the first place. A 1997 survey of Year 10 to 12 Australian high school students found that 78 per cent said that they planned to go to university for employment-related reasons, whereas noninstrumental reasons such as the appeal of interesting courses or the social and cultural benefits of the university experience figured in 3 per cent of students' reasoning (Gibson and Hatherall, 1997:125). It would appear from this study and from much anecdotal evidence that a fundamental psychological change has followed the institutional changes regarding how government, industry and students perceive themselves. An overarching ideological agenda in the "Dawkins Revolution," and one that finds clear echoes in most other OECD countries, was the creation of a "clever country": one that could excel in science and technology and feed the growing high-technology industries with new talent and innovative human capital. This has not occurred. Government, industry and universities have all been ham-fisted in their attempts to read market signals. The result is that year after year cohorts of newly qualified students, in a whole range of disciplines, emerge from university to find that there are not enough jobs to go round in their chosen profession. For many, the university experience was indeed a hiatus between high school and the dole, or if not the dole, then a job for which their university degree is either too much—or irrelevant. What is more, they will have been saddled with a debt running to thousands of dollars for the experience.

By any standard, the commercialization, informationization and marketization of the university as an institution has been somewhat less than an unqualified success—in Australia or elsewhere. Broadly speaking, what universities have been forced to change into is something that reflects neoliberal globalization; something that is inherently unstable and unpredictable, and that something, worryingly, is the primary institution for new knowledge production in the twenty-first century. The new university like the corporations involved in the globalization of capitalism that they have been forced to emulate, are on a rocky road to the future. They will need to be in a state of perpetual motion, constantly changing, reorganizing, reprioritizing and restructuring in reaction to and/or in anticipation of market signals and unpredictable policy directives. The institution has changed and mirrors society and the broad economic, cultural and political forces that shape it. The "bottom line" is all—or nearly all—and so much has fallen by the wayside or will fail to evolve and develop because of this overarching economic imperative. We shall concentrate on what we have lost and the effects of this shortly. But in closing this chapter we still need to consider that which is the most important aspect of these social cultural and economic dynamics—the students, those who will go out and help shape the primary features of tomorrow's world. What about the person who goes through the Enterprise University—and comes out the other end? We have seen something of the new student motivation—a job at the end of it—but they are met by unemployment, underemployment and debt instead. This outcome itself is unparalleled in the university experience. To appreciate the depth of these changes it may be useful to indulge in something of a comparative study between today's "typical" student and the student of the two or three decades immediately after the Second World War. What are some of the salient differences between the student of the classic baby-boomer cohorts and those of today?

It is, of course, not possible to exactly define the typical student at university at any time in history. The exercise would necessarily be a subjective one with all the biases and methodological and interpretive flaws that stem from never being able to have access to the impossible amount of data that would be needed to sustain such a project. What *can* be usefully discussed in general terms are the institutional settings of the post-war universities, the social attitudes toward them, their position *vis-à-vis* government and industry and the sorts of people who may be expected to go to them. From these broad generalizations it is possible to sketch, with a fair degree of intuitive recognition (with all its subjectivity and biases), a rough picture of what a "student in the post-war age of Fordism" in the period from, say, 1950 to 1975 was like—and then compare that outline with one of what may be called "the student in the Networked Society."

The Student in the Postwar Age of Fordism

There is no doubt that the post-war higher education system in the West was still very much an elite institutional arrangement. In what was largely a fee-paying system, only a small minority of high school students expected to go on toward higher education, and an even smaller fraction went on to complete a higher degree. *Social expectation,* as opposed to economic necessity, was a major feature of this process. Social class in Western societies was rather more rigid and definable in the middle of the twentieth century than it was at its end. Sons and daughters of middle- and upper-class families expected (and it was expected of them) to go on into higher education. Provision was made for a tiny number of "outstanding" working-class students to go on to university through scholarships, bursaries and so on. In general the student of higher education tended to be male, and until the 1960s and the growing power of feminism and the women's movement, females as a proportion of the population (or of the relevant class strata) were hugely underrepresented in universities.

Higher education systems also reflected the primary imperatives of the economies and cultures of which they were part. Under the régime of Fordism that dominated in most developed countries during this period, industrialism, mass production and mass consumption were the fundamental organizing principles. Most members in these "mass societies" were involved in the production process through factory work, trade and craft work and in general services such as transport, communications and so on. The work of the "professionals" such as lawyers, doctors, managers, engineers, architects, etc., was the preserve of the university educated and, necessarily, comprised of the minority of professionally educated who were qualified for such work. And so, reflecting as it did the primary economic, social and cultural organization of society at the time, it was impossible for a university education to be anything other than an elite pursuit. As an elite pursuit, being a student at a university was thus seen as special and therefore produced "special" people—that is, those who would be involved in the future running of society and its political, economic, cultural and intellectual institutions. A broad-based liberal arts education was seen to be the optimum preparation for future leadership, following in the ancient tradition that began with Plato's Academy. Indeed, for much of the university's existence, what a university education meant for a student *was* a liberal arts education. This was not simply an education in the humanities. It was more an *ethos* that was defined not by its subject matter but by the institution's "intellectual virtues, skills and values that permeate it and students acquired through it" (Langtry, 2000: 89). In other words, the broad-based, professionally disinterested liberal *attitude* pervaded the territory, whether one was studying medicine or science or philosophy or history. The student, through his or her academic experience, to some degree "acquired" this liberal aspect to his or her education either formally or informally. Part of this was again due to social and cultural expectations. As a "professional," a doctor, a physicist, an architect or manager would also be presumed to be conversant, more or less, with the canons of Western politics, literature or art, or have some knowledge of history, languages or philosophy. In such a *milieu*, it would have been socially debilitating (as well as embarrassing at cocktail parties) to be seen as simply a "professional" and largely ignorant of cultural, intellectual and political matters. The social organization of the university in its clubs, associations, debating societies and traditions as well as the ethos that most universities tried to sustain and the atmosphere of intellectual inquiry they tried to foster, would have made it difficult for most students to avoid non-instrumental forms of education.

At the same time it is important not to romanticize university life at this time as a place of benign elitism, where students "acquired," through osmosis, Platonic values and virtues that turned them into well-rounded receptacles of Truth and enlightenment. Reflecting the Fordist societies of which they were part and reflecting their own traditions, universities were also hierarchical, bureaucratically rigid and had adopted Taylorist principles from industry in the attempt to become more "efficient". Intolerance and bigotry could be instilled as well as (or as part of) a classic liberal education. Moreover, having a "tradition" was seen as a mark of a "proper" university, but this also meant that change was slow and that intellectual, cultural, class and racial prejudices from earlier periods lingered on and could easily be retransmitted to successive generations. Universities, in other words, tended to be deeply insular and narrow-minded places, and the "typical" students might have "acquired" these mind-sets along with a liberal education, as they too were part of the ethos of the institution.

So far I have tried to outline—admittedly in very rough and very general terms—what a post-war university was like, the type of person it attracted, and the type of person that emerged at the other end of that experience. The university was an elite experience for society's elites. There is one more factor that made the inculcation of a liberal education, along with the biases and prejudices that sat uneasily (or perhaps in some cases too easily) alongside its work, and that is *time*. It is crucially important in our comparative discussion of the universities of yesteryear and of today to realize the importance of the fact that almost every student of every student generation except our most recent one *did not have to work*. Reflecting their class position and the fact that they (or their families) would have paid for their education, students were usually financially secure during their university years. Indeed, for most universities, the idea of a student taking a job during term time was unthinkable and would not only attract the disapproval of university authorities but could result in expulsion. The institutional line was that students were at university to learn, not to make money. Work could be had during holidays, of course, and in many countries this became a traditional way for many middle- and upper-class students to experience "character-building" manual labor in farms or in factories or on the roads in trucks or vans. But term time was for study and for the cultivation of their real life and their futures.

The time factor is a little researched but vitally important aspect in the transformation from earlier forms of university education catering basically to an elite to the Enterprise University that acts as a business-cum-training-center for industry. The former university, as we have seen, reflected much that characterized the Fordist world that surrounded it. If Fordism had indeed become a whole "way of life" in society, then this also was the reality inside the university. The rhythm of the clock dominated the university as it did in society more generally, and as a level of *predictability* that characterized Fordism, it also characterized life in the university. Probably more so. Tradition and predictability go hand in hand. And so, based on this, the university had its own temporal rhythms, and these were clock based, but they were also particular to the university's own sense of uniqueness and autonomy as an institution. Semesters, class times and class duration, long summer and longish winter breaks, all had a predictability and inevitability that seemed unchanging and destined, like the Cold War or the seasons, to go on forever. Academic subjects and courses were slow to change and develop, and academic staff could expect tenure as well as sufficient time to teach and research. These temporalities produced self-assuredness, self-confidence and a certain smugness in the institutions and in their staffs and students. Importantly, it also helped foster a relative *leisureliness* in the ways of the academy that left (and was expected to leave) time for contemplation, reflection and debate.

The student of this former manifestation of the university was undoubtedly amongst the elite and the privileged of society. Part of his or her good fortune was to experience and utilize the unique temporal rhythms of university life, learning and culture. Universities were not temporally synchronized with industry as they are today. The experience thus afforded the time, in many ways, to consolidate one's place in society through *becoming* the doctor, the engineer or the philosopher in an institution and in a social system that had its own predictabilities, certainties and sureties that marked Fordism as a social system. Depending upon how one looks at it, the university in the age of Fordism was a degraded (or improved) version of what existed earlier, that is, an even more elitist institution for an even tinier fraction of society, or one that was slowly responding to the imperatives of a mass society. What it could still unambiguously offer, though, was the *time*, the *availability* and the *opportunity*, if one so chose, of obtaining a liberal education and the central role for critical thought that this provided for.

The Student in the Networked Society

The liberal education of the kind just described is, of course, still available. Students do not even need to go to Oxford or Cambridge or one of the US Ivy League institutions to attain one. Many other universities still offer enough choice and mix of vocational and liberal arts subjects to enable the student (who would also be unencumbered by having to earn money) to emerge with a well-rounded education, with a professional expertise as well as all the "intellectual virtues, skills and values" that a liberal education may bestow. However, these "virtues, skills and values" no longer pervade the university and no longer constitute its background logic. They need to be consciously sought out or recreated from scratch. Institutional and social changes have made a liberal education much more difficult, but not impossible. However, we also live in the age where the higher education system has massively expanded to cater to the needs and imperatives of industry and the economy more generally. Citing the US experience, Christopher Lasch (1995:177) writes bitterly that the liberal education, with its emphasis upon critical thought and reflexivity

> has become the prerogative of the rich, together with a small number of students recruited from selected minorities. The great majority of college students, relegated to institutions that have given up even the pretense of a liberal education, study business, accounting, physical education, public relations, and other practical subjects. They get little training in writing (unless "Commercial English" is an acceptable substitute), seldom read a book, and graduate without exposure to history, philosophy, or literature. Their only acquaintance with the world's culture comes through required courses like "Introduction to Sociology" and "General Biology." Most of them hold part-time jobs, in any case, that leave little time for reading and reflection.

As Meaghan Morris and Iain McCalman (2000) argue, the liberal education "takes time." However, operating in an accelerated or real-time environment leaves ever-diminishing time for it. Consequently, the end of the "pretense of a liberal education" and the shift to vocationalism and commercialism has spelled disaster for the arts and humanities in the university. Important subjects which can instill the ability for critical thought such as philosophy, literature, social theory, politics, history and so on are considered to be largely irrelevant to the "bottom line." These disciplines have suffered grievously in the change to the enterprise model. Students in the main see no future in the study of these subjects, and academics who teach them feel vulnerable and marginalized, as they see dwindling research funding for programs that both industry and government regard as a luxury we can ill afford. As a consequence our social capital of critical thought has begun to diminish precipitously.

Lasch's "great majority" now move through from university to industry, become the shapers of the future, the leaders in business, commerce, management, culture industries and the universities and will be the carriers of society's moral, social and cultural values. Unless there are some drastic changes in the ways universities function, their huge preponderance in numbers and their vocational, instrumental skills, so necessary for a society reconstructed to reflect the needs of the economic "bottom line," ensure this. In this mass cohort the majority of students formerly would have been workers, with no necessary interest in higher education as it once was—or was idealized to be. Higher education has become a means to an end—the qualification necessary (or hoped to be necessary) for a job that (students hope) may exist in three or four years time.

I have tried to show that the university was once a place with its own temporal rhythms. These were predictable, possibly too predictable and narrow-minded and too slow to change, but they did at least have a measure of certainty. Universities are now synchronizing to the temporal rhythm of industry, and this rhythm is real-time and chronoscopic. Aping corporations in the marketplace, as they now do, universities operate in a climate of perpetual uncertainty. Staff fear for their jobs; departments endure constant reorganization and reprioritization, and the continual need to attract alternative sources of funding leaves little time for research or commitment to teaching. Students end up confused at the constant changes within their university and cynical or disappointed regarding the level of "service" they get. The students' lot is compounded by the fact that many of them arrive at the informationized and commercialized university with little idea of what they really want to do. They have one eye on how the job market is shaping up (as does the university), and they try to anticipate its needs. They then go for what they hope is the necessary skill to qualify them—or at least put them in the reckoning—alongside the many others who are making the same strategic judgments. The focus of the other eye is switching between the needs of surviving whilst studying, that is, looking for or working at a job, and the tasks supposedly at hand—going to lectures, reading, writing, doing research, debating, discussing and questioning with teachers and fellow students, and so on. Increasing numbers of students view actual study as simply another obstacle to be overcome by whatever means possible—like finding a job and picking the "right" course. Recent studies have suggested that the reorientation of the university into an enterprise model that seeks validation from the marketplace and from its "customers," that is, the students, has created a "culture of entitlement" where students get higher grades for less effort and where fewer students believe the university experience has any effect upon their personal lives. Craig McInnis from the Centre for the Study of Higher Education at the University of Melbourne in Australia argues that students are now far more pragmatic about why they are at university and what they need to do to "succeed" there. McInnis (2001) noted that:

> The range of institutions, courses and subjects now available, combined with the increasingly sophisticated access to flexible modes of knowledge delivery and electronically generated communities of learners, puts students in a powerful position to shape the undergraduate experience to suit their timetables and priorities.

The unintended "empowerment" of the student has paradoxically led in many cases to rising levels of uncertainty, confusion, cynicism and an instrumental attitude toward higher education. Being in a "powerful position to shape the undergraduate experience" necessarily devalues this experience when other, less controllable forces such as work and time pressure impinge constantly upon the learning process. Moreover, McInnis asserts that a poor understanding of the changing nature of the new student makes universities likely to accept improvised solutions; they have also become excessively responsive to what students think they want (what they have paid for) as opposed to what may benefit them as individuals and as members of society more generally. An example of this excessive responsiveness may be seen in the University of Virginia's Virtual University, which encourages students to create *their own* curricula. Dana Mulhauser explains the now-predictable logic and values:

> The university would act as a broker, helping students develop personalized curriculum from resources at both in-state and out-of-state institutions, as well as from businesses such as Microsoft.... Tuition would vary by student and by semester, based on the costs of the online classes in which the student enrolled. Counseling by the university's faculty members would carry a per-hour fee. (Mulhauser, 2001)

Some quantitative measure of the new logic and values may be illustrative of how much things have changed. For example, a study in the USA that drew upon survey data from 9 million freshmen over the last thirty years has concluded that there has been a major shift in student values. Asked whether a primary motivation for going to university was to "develop a meaningful philosophy of life" or to be "well-off financially," the study found that the two values have switched places over the last thirty years with financial well-being as the central value for 74 per cent of students (Astin, 1998). Moreover, the study reported a marked decrease in the number of hours students spend per week in study, an increase in the numbers who miss classes, less time spent with teachers, and an increasing worry about finance and how to make ends meet. Students also feel more overwhelmed by all that is expected of them, are more materialistic and expect the university to help them achieve their more instrumental goals.

Such is the form and function of the Enterprise University that has replaced the languid, self-confident, predictability of the university education for the elites. I am not here calling for a return to those times— far from it—but it has to be recognized that the new university and the "clients" or "customers" it "processes" are in no way preferable. Indeed, as I shall argue later, in terms of a functioning of a vibrant and pluralistic civil society, they may be infinitely worse.

Working and Learning in Real-Time

There is no doubt that the university student of today is a vastly different creature from that of even a generation ago. The world of work and the world of learning have come together in ways that would have previously been unthinkable to government, to industry, to the students and to the university itself. In an accelerated world, students exist in real-time, and this means a constant juggling of the priorities of work and study, social life and family life. As an article by Bruce Horovitz in *USA Today* (1999) article puts it:

> You apply. You get accepted. You buy a futon. You register for classes. You gripe about dorm food. You go to class. You study. You order pizza. You listen to music. You break up. You argue with your professor. You sell your books for beer money. You graduate. All online.

But of course it is more than this. Far from freeing one to use time efficiently and possibly saving time for other things, a deepening interconnectivity based on industrial chronoscopic time merely stretches the individual to the limit in terms of what he or she is able to do and continually acts as a vector for yet more demands upon one's time. This increasing interconnectivity is one of the dominant features of our era. It is on course to permeate life more and more comprehensively as competition and evolving technological breakthroughs breach the barriers of private life and of non-commercial forms of social and cultural interaction between individuals and societies. In this real-time, time-pressured environment, nothing gets done properly. Society is geared to the temporality of the network, and society's institutions work in tandem to keep the rhythm up. It was noted previously that universities used to take a dim view of the student who worked. A *Times Higher Education Supplement* survey conducted in British universities in 1998 noted "nearly every university in the country runs or supports a service to help students find a part-time job while they study" (Swain, 1998:2). In the USA surveys have reported the significant impact of paid work upon study. In 1999 the

American Council on Education found that approximately four-fifths of students in higher education were working whilst enrolled in an undergraduate degree. Twenty-seven per cent were working between twenty-one and thirty-four hours per week, and around one in five was working the equivalent of a full-time job (McInnis, 2001:6)

And in the time-pressured information ecology, this has a knock-on effect elsewhere. Studies have found that students' jobs impact upon their academic achievement. It is a complex area to measure, but it seems that the consequences are negative, especially the longer the student works at a job during the week. Sometimes the work-study nexus can on the surface look positive, but on closer inspection it may be that when a high percentage of students in a class hold part-time or sometimes full-time jobs, the learning environment changes because teachers' expectations of student performance begin to lower (Gehring, 2000). McInnis (2001:4) sees these trends as representative of a generalized student "disengagement" from the university experience. His findings in a 2000 national trend survey in Australia are worth quoting at some length:

> Students are spending less time on campus and more time working in paid employment. They seek, or have in their lives, an increasing number of activities and priorities that compete with the demands of the university. Aside from the growing impact of part-time work painfully obvious to academics, students have less need to spend time on campus in order to study, or to have access to teaching and learning resources. However, there is also evidence of a declining level of student commitment to university study that is not entirely explained by financial pressure on individuals or the availability of information technology. There has been, for example, an increase in the proportion of students who say they find it difficult to get themselves motivated to study, and also in the number of students finding the workload difficult to manage. A related trend concerns crucial aspects of study habits and social learning opportunities. For example, students are less likely to study on weekends, and are more likely to frequently rely on friends for course materials.

A related trend, though for obvious reasons a more anecdotal one, is that of student plagiarism. Of course some students have cheated and stolen their way to a degree for a very long time. And universities, rhetorically at least, still rate plagiarism as the most heinous of intellectual crimes. But it is nevertheless on the increase and is quite possibly rampant in most if not all higher education establishments. The proliferation of Internet sites that sell readymade essays and the rapid development of software to detect plagiarized essays are testimony to this. Quite simply, the university that sees itself as a business and is addicted to ICTs, coupled with students who are increasingly "disengaged" from university life and its erstwhile values, is set up for it. Plagiarism is easy using ICTs, and resort to it is becoming irresistible for increasing numbers of students. A study carried out by Steve Jones (2002a) for the Pew Internet & American Life Project, entitled *The Internet Goes to College* found that 73 per cent of college students use the Internet more than the library. This in itself does not indicate plagiarism, of course, but it is part of a logic-shift away from books and reading toward surfing and the temptation to cut-and-paste a research paper, a temptation made all the more irresistible in a time-starved students life. Some idea of the scale of the problem can be ascertained in the case of the University of California at Berkeley professor who caught 45 of his 320 neurobiology students through use of a software program that scans the Internet for "matches" (Schevitz, 1999). This probably understates the level of plagiarism, as the program will only recognize the more blatant and egregious attempts at intellectual dishonesty. Cut-and-paste work may easily be "hidden" from such programs through a variety of methods such as paraphrasing, changing key words by using the word-processing thesaurus and so on. Besides, many teachers will simply not have the time to devote to such depressing detective work. They know that some students cheat but then will reason that it is too hard to catch them or that the resulting disciplinary procedures and hearings, appeals, etc. will be too messy and time-consuming, and that this is a path they do not really want to go down.

Teachers may be implicated in catering to the needs of students in other ways. In the Enterprise University, where non-government sources of funding are becoming more and more important, there is pressure to keep fee-paying students from failing. After all, telling a student that he or she hasn't made the grade this term, potentially, means cutting off much-needed revenue for a department. Failing two or three students could mean the loss of thousands of dollars to a cash-strapped faculty. Such pressures have led to a spate of allegations over the years of so-called "soft-marking," where implicit or explicit pressure is applied to upgrade students' work to allow them to pass and make them eligible for further study for a higher degree. In 2001, Ted Steele, a science professor at the University of Wollongong in Australia, was sacked because he went public, alleging he had been told to upgrade student work. Like plagiarism, this is a grey area where no one usually speaks candidly unless they are "whistleblowers" such as Steele and are prepared to put their job and career on the line. The reality is a subterranean world of self-censorship and self-delusion, where substandard students pass and professors keep quiet or convince themselves that the student really did deserve the grade—or a "helping hand." As for the students themselves, in the "culture of entitlement" increasing numbers of them are of the opinion that if they have paid for a service, they expect to get it.

This short comparative exercise has illustrated the stark contrast between universities and students as they were not so very long ago, and how they are today. As institutions, universities have radically reoriented themselves to the needs of the economy and have become primarily vocational forcing-houses for industry. In itself this is not wholly bad. But the cost has been the neglect and decline of a liberal education and the tremendous value that this has for society. Students, in turn, have changed in their attitudes as to what a university is for. It holds no special place in their lives any longer—even if that special place was just as a mark of status. It competes with other parts of their lives such as work, socializing and family—and often comes well down the list. Learning has become an instrumental task on the road to (students hope) a well-paid job. Indeed, as we have seen, learning is not the goal for most students any longer—getting the credential is—and for the pragmatic student in the networked society, any way of doing this can now be considered. The next section will deal with the process of learning—still the official rationale for most universities. How does this work in the chronoscopic-networked society, with all the attendant time pressures that come with it?

Chapter Eight
Temporal Sense Making in the Information Ecology

A central concern for much of this book, indeed the intention behind the first five chapters of Part One, was to explain the importance of *time* in industrial society and the subsequent shift from a chronologic to chronoscopic temporality through the impact of the globalization and information technology revolution. In the present chapter I want to discuss the process of *learning* and *knowledge production* within this information ecology—a realm of inquiry where almost no basic research has been done—and yet is the direction in which society is heading in pell-mell fashion. This temporality is much more than the "accelerated" pace of life that Gleick observes and almost all of us feel, where we constantly are starved of time, feel harassed and harried about the demands made upon our time and must endure the feelings of frustration and incompleteness in our lives and our daily life's tasks and projects. These are indeed realities, but their underpinnings go much deeper. I am here arguing that through the impact of globalization and the information technology revolution, we stand (if "stand" is the right word) at the beginning of a process where revolutionary technological change is spelling the end of time, as we have known it for a thousand years. The meter of chronologic clock time that has underscored our industrial, social and cultural institutions, and has been the temporality through which we have made sense of the world, is being supplanted by a digitally compressed temporality that operates in a real-time of constant duration. Through the increasing density of data networks and human interconnectivity we are in the process of creating a whole new temporal ecology based on the constant *now*. Inside this "buzz of the flickering present" we now try to orient ourselves within a very new world; a cyberspatial world based upon a new machine time, where seconds have been concentrated into nanoseconds, and like the bullet that rips through the balloon in super-fast photography, we are suspended in the constant present of the information ecology indefinitely. Like the chronologic world that began to emerge in medieval times, the chronoscopic world is a social construction—and our worldviews, the ways in which we think, see, and reason now begin to reflect this new temporality.

I want now to discuss how these functions and processes of learning and knowledge production are being shaped and changed by the radically new information and temporal environment in which they now take place. I shall begin this discussion on temporal sense making with another short scenario. It is one that will be wholly recognizable to millions of students today—and would be totally alien to most students of a generation ago. In other words, it characterizes the radically new temporal life-world of students (not simply as students, but as workers, too) in the age of globalization and the networked society.

Scenario: A Day In Wired Life

Wednesday. Phillip wakes up at 7:15, tired, he was working until late the previous night for a telemarketing company located in a major call-center in town. After several frustrating attempts, he eventually manages to log on to his university website and downloads the lecturer's PowerPoint presentation from the Politics 210 class yesterday afternoon. He can't make it to that five o'clock class because he begins work at six on Tuesdays. His dad was a union organizer and had always tried to instill in Phillip an interest in politics, and so he tries to take an interest. He curses over the PowerPoint notes as he scans them. They don't make much sense—they never do—they're comprehensible only to the lecturer who extemporizes from them, because he knows their context. It's nine o'clock already and he has a Marketing 202 lecture at eleven. Marketing is where he wants to work. Not because of any particular interest in selling, but because that is where the jobs are these days, apparently. Phillip logs on again and emails his Politics lecturer to ask if he can give him a summary of the talk—and maybe explain what he means by Fukuyama's concept of "Trust" and give some examples of the supposed "crisis of trust" he refers to—he may try to use this as a research topic for his major essay.

He checks his email in-box and finds a message from his friend Catherine who is usually able to make it to the Politics lectures, though not Marketing 202, so they have a sort of information-swapping arrangement. She tells him that as of next week, Politics 210 lectures will be available through online video streaming. Problem solved, he thinks, never need to miss another class. Phillip emails the lecturer once more to ask if this is true, and could he give him the website address. His cell-phone rings. A text-message from his co-student and fellow-worker, Ian, informing him in SMS shorthand that the boss was looking for him just after he had left work last night. Phillip replies with a text-message to the effect that he wasn't to tell the boss he's seen him. Phillip was still online, so he decided to check the Internet for jobs in marketing, then got distracted by a website devoted to his favorite football team, then got involved in a chat room discussion of the game he missed last night. He suddenly realized he was running late and would have to rush to get to the eleven o'clock class. He hadn't been able to do any of the reading this week due to working at his other job as an unpaid intern at a marketing company. The university got it for him and it's only seven hours a week, but he needs the work experience and the contacts.

Phillip runs up the platform ramp at the railway station and manages to get on board just as his train was due to leave. He should make class easily, with half an hour to spare even—time enough for some speed-reading of this week's marketing downloads. His cell-phone rings. It is his boss, Steve. Can he come in for a few hours over lunchtime, twelve until three? Sandra called in sick last night, and there's no one else available. Phillip knows this is not a request from Steve but an order, and if he says no, then he can forget about getting the amount of hours he needs to survive. In fact, make a habit of refusing to comply and he can forget about the job, period. Sure, he tells Steve, it'd be a pleasure to help out—just let me know, anytime. I'm a team player. He phones Catherine and tells her that Marketing 202 is off the list of priorities for the day and that he's sorry she'll have to miss out on his notes, however thin and incomprehensible she usually accuses them of being. Catherine says that a student she knows called Janice always has Wednesdays free from work, and she is sure to be there—moreover, she takes a laptop to class for note taking, and emails her notes to Catherine. She's not all that interested in marketing, and got terrible grades last year, and the notes are practically useless, but would he like her email address anyway?

Phillip's three hours at the call-center ran predictably to six, and by the time he got home it was 6.30. He calls Sandra and asks if she is going to be off sick for much longer—another week the doctor said, Sandra replied. Phillip muses whether to switch off his cell-phone in case Steve rings again tomorrow; he really should try to get a clear day for reading.

He then decides that a switched-off cell-phone would be construed by Steve as a refusal and a sign of not being the mandatory team player. It'll have to stay on. He gets online and downloads the Marketing and Politics lecture notes for next week and makes a check of the library catalogue to see if the relevant readings are in—they are. He makes a mental note to try to get to the library over the weekend, in between the call-center and the job he has had since high school at the hardware store on Sunday afternoons. He emails Janice, introduces himself, and asks if he can have the lecture notes as an email attachment. Phillip stays online and checks out the marketing jobs sites he has bookmarked and dreams distractedly of a well-paid position with one of the big advertising companies in the city. A pop-up box tells him he's received an email. Its from Janice who says that her laptop crashed midway through the lecture, and she lost all that she had written—maybe I can send you next week's if you have to work, she adds by way of an apology. Another email in the in-tray, this time from the Politics lecturer. In essence, he says that he is only a sessional lecturer, paid by the hour to give lectures; he doesn't have an office, and he only comes to the university to take the class. He gives Phillip a list of four readings related to his question, asks him to read them, summarize them, note down some questions on what he still doesn't understand, email all this to him, and he'll try to reply as soon as he can.

Phillip tries to let this all sink in and then looks at next week's list of readings that he has downloaded. The marketing titles go into his Google search engine and elicit lots of websites that deal with just this stuff. He scans it desultorily; he is getting tired and hungry. He bookmarks some sites of particular interest and makes another mental note to try to look at these properly when his assignments are due. Back to the search engine. He types: "Fukuyama" and "crisis of trust" and looks at the titles of the first ten of the 2,890 responses it took just 0.52 seconds to deliver up. One of them gave access to a full-text article on just his subject, but the PDF file was 45 pages long and looked like very difficult stuff—it was for a post-grad class somewhere in Canada. Phillip cut-and-pasted the author's email address, the paper's abstract and downloaded them into his PDA with a note to himself to email the author with a list of questions when he was beginning the research paper. Maybe she'd explain it to him. If she's full-time, maybe she'll have the time to do it.

It is getting late. He thinks vaguely about a Marketing paper that is due in about ten days time. A guy at the telemarketing company who seems friendly and knowledgeable might be worth a quick email or phone-call. He may even have written something on that question himself; he did the same course two years ago. Phillip then remembers that tonight he is supposed to take part in an online game of *Resident Evil*; he'd bought a new XBOX and is keen to try it out. Getting everything set up took about half an hour, and it's close to ten now. The cell-phone rings. Phillip looks at the number on the LED screen and sees that it is Steve. He lets it ring and go on to voicemail. A few minutes later he checks the voicemail and hears Steve asking him to come in for a whole day instead of the half-day he usually works on Thursdays. Sandra is still sick, he says; it sounds like she may be off for a month.

◫ ◫ ◫

It is worth repeating: this small scenario is by no means outlandish or exaggerated. As both a student and teacher I have encountered many scenarios such as this either personally or through colleagues or fellow students. Many readers will instantly recognize some of what is included in this little vignette. What is important is to also recognize the *pervasiveness* of such relationships in what is supposed to be an environment of learning, inquiry and reflection. The Phillip character in the vignette would not even be classed as a "disengaged" student of the kind McInnis has found in his studies; he is one who tries and battles with frustration and incompleteness. The "disengaged" student hardly cares any longer, the "pass" mark is all he or she wants, and as we have seen, as customers, they come to expect at the very least not to fail—they've paid, after all. Supposing Phillip began to view his Politics' lectures, maybe in time, *all* his lectures on video streaming, at his leisure, when he had the available time? A solution, perhaps? Harvard University offers just such a service to its students, but Harvard professors complain that the video lecture encourages students to skip classes and cram online prior to tests. The Harvard Instructional Computing Group estimates that 70 per cent of those who access online lectures watch between 5 minutes and 30 minutes, indicating a "skimming" of the video, selecting just those parts deemed necessary.

It is more than just delivery, though. It is about living and earning (and trying to learn) within the information ecology created by the globalization/ICT nexus. The economy is not synchronized to a temporality that allows for learning, and the universities have turned their backs on a major part of what they used to be about: the inculcation of a capacity for critical thought. This takes time and in the chronoscopic temporality, there simply is no time for it. The real-life Phillips, Catherines and Janices do learn, but they learn to develop pragmatic coping strategies such as "time management" that evolve and are honed to achieve the optimal outcome with as little "collateral damage" as possible to other facets of a time-pressured life.

Time for Learning?

It would be fairly uncontroversial to assert that there is a temporal element in the process of learning, of knowledge production and knowledge acquisition. Notwithstanding the different theories of learning and cognition, the time factor is something that is at least implied as necessary in any theory of learning. The very terms such as "practice," "learning," "knowledge" connote a temporality, a time duration that is of some length. On a common-sense level, one "knows" intuitively that learning takes time and cannot and does not come rapidly. For example, think of reading, and having to read quickly, a 5,000-word essay on, say, a new theory of jurisprudence on the international law of the sea, something I know virtually nothing about. For me, this *literally* would be a waste of time. There would be nothing I could connect this to in a wider body of reading and knowledge that I would have to be familiar with for these 5,000 words to make sense and to be part of a new knowledge acquisition. From these 5,000 words I would need to work backwards to the theories and issues and histories these words make reference to and have evolved from; I'd need to work outwards to the contemporary debates that inform these words and give them context, and I'd need to think forwards to ask if these 5,000 words constitute a worthwhile addition to knowledge in this area. Only then would these 5,000 words have the vaguest of context and sensibility—and this would be after many hours of reading and thinking and talking with others in that field of knowledge. Moreover, this process has nothing to do with the delivery of the material. It could be online, video streaming, paper-based or even an oral delivery. The cognitive process itself, from words on a paper or sound and vision to knowledge and understanding, has a temporal dimension. Indeed it is not fanciful to speculate that learning, cognition, understanding, whatever we wish to call it, has its own biologically based temporal rhythm in the same ways as a woman's reproductive cycle or the circadian rhythms that reside in us at the molecular level.

The Mechanics of Knowledge Production

As I have just intimated, theories of learning are not the issue—lack of time and the chronoscopic environment in which learning now takes place are. Nevertheless, there are broad parameters, contexts, processes that need to be discussed, and these are the mechanics of knowledge production. What actually drives knowledge production, and what knowledge is, is a problem that is central to philosophy or, more precisely, its sub-branch, epistemology. It is a problem that goes back to Socrates, through to Descartes, Kant, Hume and Spinoza in the Age of Enlightenment, and more recently through the works of philosophers of the episteme such as Wittgenstein, Foucault and Lyotard. Here is not the time or the place to analyze these strands of philosophic thought, the merits of empiricism over rationalism, or vice-versa. Nor can we discuss equally fine-grained and critical questions such as: What *is* knowledge? How can we be certain of our knowledge? Upon what basis do we evaluate claims to knowledge? The philosophical questions are important, and much more research still needs to be done, as the validity of these questions will not diminish, indeed it will only increase, as the network society becomes more all encompassing of culture and society. This is especially so in the context of the university, an institution that is still pivotal in the creation and the dissemination of knowledge in society. However, to get at the essence of the dynamics affecting the production of knowledge today, to try to capture the mechanics of knowledge production in our accelerated society, I want to broach these questions from the perspective of the sociology of knowledge. In particular I want to discuss and critique two important books that have theorized the changes in the production of knowledge in the universities over the last twenty years.

The first is *The New Production of Knowledge*, written by Michael Gibbons et al. in 1994. This book was never given anywhere near the attention it deserves as one of this first to look at the impact that the radical restructuring of the economy and society in the West was having upon the universities and the knowledges they produced. The authors identify two distinct modes of knowledge, the differences between which represent an emerging trend in the kind of knowledge that is being produced and their effects upon the economy and society. What they call Mode 1 constitutes the traditional, discipline-based forms of knowledge that are delineated into basic "theoretical" and "applied" forms. Mode 1 knowledge evolved directly out of the Newtonian-based principles that govern what is considered legitimate "science." This was the basis of the knowledge production that underpinned and sustained Enlightenment-based modernity and capitalist industrialization. Mode 1 knowledge production was characterized by its hierarchy and its homogeneity and was legitimated through its sole site of production—the university (the Fordist post-war university I sketched previously, for example). Problems and issues were recognized and solved by the primarily academic interests of a specific discipline-based community. If it didn't come from the academe, in other words, it wasn't a recognized problem; it wasn't science and it wasn't knowledge.

Mode 2 knowledge production represents a shift from this rather rigid and Fordist method to one that may be described as post-Fordist and *flexible*. It responds to practical problems and issues arising in and identified by actors in society more generally, not only in the university. Indeed, the authors argue that in the near future "the universities, in particular, will comprise only a part, perhaps only a small part, of the knowledge producing sector" (1994:85). The authors argue that this is indeed a good thing and that the diffusion of knowledge production out from the universities and into society more generally may lead to a "democratization of knowledge." This new form of knowledge production, Gibbons et al. maintain, has been brought about through the evolution of "the knowledge society," where "knowledge workers," across many and varied industries, comprise the momentum behind the post-industrial society. And in a vast and increasingly complex society, practical necessity has meant that the disciplinary boundaries have become blurred, and the sites of knowledge production have proliferated—out of the university and into industry laboratories, independent think tanks, workers engaged on the job in solving practical problems, and so on. The most important aspect of Mode 2 knowledge production in the post-industrial world, the authors argue, is that the dominant forms of knowledge being produced are rational, instrumental and characterized by a specific problem- and goal-oriented focus.

In his 2001 book *Challenging Knowledge: The University in the Knowledge Society*, Gerard Delanty tries to build upon the insights offered by Gibbons et al. Whilst agreeing with the book in the broadest of terms, that is, that there has been a major "epistemic/cognitive shift" in knowledge production, he differs significantly about what this means. For instance, Delanty acknowledges that there has been a definite shift toward the instrumentalization of knowledge. He insists, rightly I think, that this is highly problematic and that knowledge is rather more than a matter of "expertise" (p. 5). Moreover, he is skeptical, again rightly, of Gibbons et al. claim that the diffusion of knowledge production sites out from the university and into the private industry labs and think tanks represents a democratization of knowledge. Delanty also takes issue with the implied unproblematic "decline" of the university as key knowledge producer. He acknowledges the extent of the deterioration in the shift from modernity to postmodernity but argues that the university has to reinvent itself and must forge for itself a new social, cultural and intellectual role. Key to Delanty's thesis is that with the supersession of Mode 1 type knowledge production, new knowledge production in its transdisciplinary diversity has become more reflexive, in a way, released from the rigidities of modernity. He argues that this "reflexive application" stems from the fact that "knowledge is being used to produce knowledge and the conditions of knowledge production are no longer controlled by the mode of knowledge itself" (2001:152). It is in the "reflexive application" of knowledge that the university can reinvent itself and its role in society; in effect, to critique and extend the limits of knowledges being produced elsewhere, acting as a space "where different discourses" and different epistemes "interconnect" to form fruitful communities of knowledge (p.151).

It would be clear to the reader that I would agree with both Gibbons et al. and Delanty that there has indeed been an "epistemic/cognitive shift" in knowledge production, one that stems from an agreed source, that is, the shift that has been labeled as that from modernity to postmodernity, from Fordism to post-Fordism, or from industrialism to postindustrialism. I agree also with the contention in both books that there has been a profound shift toward the production of instrumental-rational forms of knowledge. What is far from clear, however, is how the diffusion, "the democratization" of knowledge characterized by the breakdown of Mode 1 knowledge production, ipso facto represents the basis for "reflexivity" or the ability of knowledge to critique itself. Logic *would* imply that if knowledge production has become less rigid in its sources of legitimation, then "spaces" for reflexivity could, in theory, be able to flourish. However, this overlooks some major influences that this book has been concerned with.

First, if the agreed emphasis has been toward more instrumental forms of knowledge production, based in private laboratories and privately endowed think tanks—as well as in the "Enterprise" universities—then from where do the "spaces" for reflexivity emerge? It seems to me that although the rigid, prescriptive modes of what constitutes knowledge in Mode 1 may have largely broken down, the new commercialized mode is flexible in one sense, that is, it is flexible in that it is able to respond rapidly to the commercialized and commodified problems and challenges raised in the neoliberal age of globalization; and indeed, it may also be reflexive, able to think deeply and critically about such problems and challenges as and when required. However, Mode 2 is also extremely rigid, I believe fatally so, in that the scope of this knowledge is limited by that which is relevant to the economy and is legitimized by market-based criteria (Lyotard, 1979). Who will pay for critique and reflexivity when what the market demands are answers and results? Where will it stem from otherwise, in modes that are effective, powerful and coherent enough to have an influence upon culture and society?

Second, not enough emphasis is given to the role played by the nexus between neoliberal globalization and the ICT revolution. How are reflexive forms of knowledge able to grow and critique other knowledges in the network society, where the commodificationary impulses and the temporal acceleration within its domain favor instrumental thinking and directly militate against reflexivity and critique? Temporal acceleration and the imperatives of the market are what sustain the network society, a society that is insinuating itself more and more into realms of culture and our civic institutions. "Spaces" for reflexivity, as a direct consequence, are being marginalized and their effectivity severely truncated.

The problem, I believe, is that Gibbons and Delanty look at knowledge in specific way. For Delanty, especially, it seems to be simply a case of the end of the rigid Mode 1 model allowing the thousand flowers of reflexivity to bloom. This is a consequence, perhaps, of his method and intellectual approach. He looks at the works of many of the major social theorists of knowledge, many of them dead, and gives very little attention to the effects of neoliberal globalization and, vitally, the ICT revolution. Neoliberal globalization and the ICT revolution have, I believe, had the effect of augmenting, massively, instrumental forms of knowledge within the information ecology, at the expense of all others that do not fit with its aims and values. Moreover, the processes of temporal acceleration within this ecology simply leave no space and no time for reflexivity to grow and have an effect.

What I want to do now is to uncover the mechanics of knowledge production in a way that is as clear and succinct as possible. In so doing I shall endeavor to make *explicit* that which is *implicit* in their processes, that is, the centrality of time to the sustenance of critique. What follows is a fairly simple and non-contentious step-by-step guide through the stages of what we may call "the getting to knowledge." Knowledge, however, is not here taken as an *a priori* given. As we shall see its shape and content, its forms and substance, are "socially sustained and backed by power" as Ronald Barnett puts it (1997:5). In other words, like time, knowledge is a social construction. Here I have drawn liberally upon Charles Jonscher's *The Evolution of Wired Life* (1999), because he deals with such a potentially rarefied subject in an intelligent and readable way within the context of the network society, and it would be futile for me to attempt to improve upon it. Jonscher describes three phases in the production of knowledge. These are: (1) *Data* (2) *Information* and (3) *Knowledge*.

Data

The word "data" comes from the Latin meaning "things given" and in this context refers to the signals and symbols that confront us in raw form in the world around us as they act upon our five primary senses. The most efficient data collectors amongst our senses are the eyes and ears because they work by detecting oscillations in the environment, primarily light and sound signals. Our eyes, for example, detect with great acuity the different frequencies present in light, perceiving them as different colors. When these rays of light reach our eye, the cornea focuses them onto the retina. These light signals are processed into an image on the receiving sensors in the eyes. These sensors comprise over 120 million photoreceptors in the retina, called rods and cones. The rods are receptive to light and dark changes, shape and movement. The cones are not as sensitive to light as the rods, but are sensitive to one of three different colors (green, red or blue). The amount of light signals generated by the environment around us is, fairly obviously, almost limitless, far greater than the eye's photoreceptors can ever hope to assimilate. In this raw, "given" form it is simply a vast mass of data with no intrinsic meaning. To the "untrained eye" the world would seem a chaos of shapes and colors. We first have to attribute meaning to this raw data. From where does meaning come? Robert Hughes, in his 1986 book, *The Fatal Shore*, has described how this works in reality. He tells the story of the British ships that arrived in Australia in 1770 at a place they called Cape Everard. The Aborigine fishermen in the bay "saw" the ships approaching them but went on fishing, seemingly oblivious to their presence. The problem was that the height and shape of these ships were so fantastically alien to these indigenous peoples' frames of reference that they simply ignored them. They possessed no adequate way of response. In other words they could see the ships, obviously, but they held absolutely no meaning. However, if they had seen these ships, or similar ships before, then these shapes would have some meaning, some frame of reference, however tenuous or sketchy, in which case the raw data that comprised the ship or ships could be processed into *information*.

Information

The root of the word again is Latin, *informare*, and means "to describe." Such information is a description, a statement of data in the abstract. On this abstract level, the movement or shapes in which the data represents it is processed by the primary cortex region in the brain into information that has a level of meaning. Roughly twenty billion neurons residing in the visual cortex interpret (process) the data coming in through the optic nerve. These neurons act upon the ways in which we mentally and physically respond to the data. The next stage is to favor the center of the image. We see surprising little of what is in front of us. Developmental psychologists, for example, have studied the way very young babies will grasp at an object suspended in front of them. They will begin by flailing randomly at the object, trying to grasp it. Photographic evidence shows that the baby will try to grasp the object from many differing angles. However, once the baby succeeds in learning how to use hand-eye movement to focus on the center of the object, it will succeed in grasping the object and once it has had practice in doing this, over time the "wasted" grasping movements will stop and the baby will aim straight for the object. Researchers believe that what is happening is that when the grasping action is successful, the firing patterns of the neurons in the brain at that time "encode" that successful pattern onto the brain and will trigger a processing of the data into "recognition" or meaningful information for the baby if the object is seen again. Similarly, the Australian Aborigines described by Robert Hughes would have processed the alien and incomprehensible shapes of the British ships into something recognizable, after their initial exposure. Meaning could arguably be imputed, and they could have represented ghosts, sea monsters, or virtually anything that would fit their own (now gradually expanding) frames of reference. This would be meaning, yes, but not *knowledge*.

Knowledge

The arrival at a state of "knowledge" is not the same as the shift from the interpretation of a stream of photons (data) into information. It is the result of an immensely complex interaction of previously learned ideas and concepts, experiences and recognitions that are carried out in a higher-level function of the brain, in the cerebrum. The precise workings of the cerebrum are not fully understood, but it is here that consciousness is thought to reside and thinking takes place. Knowledge builds, like sedimentation, over time. The production of knowledge, both in our own heads and as a social store of knowledge in writing, in books, in films, in ancient tablets, in computer memory banks, in customs, practices, language and so on relies on the "vast reservoir of pre-existing knowledge residing in the mind, both inherited from our ancestors and based upon the experiences of our own lives to date" (Jonscher, 1999: 44). The mind is the prism for this process, and according to Kant the mind has developed a framework for "prior knowledge," a "training" and a cultivation of the thought processes which allows for meaning and ultimately knowledge to be derived from the mass of ultimately meaningless symbols and signals, "things given" that surround us in the natural world. The Germans call it *bildung*, and it is something that is rapidly becoming obsolete in the network society. This process is an intensely *social* one, where what construes knowledge and what forms of knowledge predominate in society need to be seen in the context, as Barnett noted above, of "power" and where it resides in society.

Much of what I have written thus far regarding the evolution of the university as a social institution and the broad changes in culture, economics and society that the changing university reflected was underpinned by the recognition of the growing tension between competing forms of valid knowledge in society. We saw the dualism between theological "truth" and neo-Platonic metaphysics in the Middle Ages, through to the long phase of tension between, on one hand, utilitarian and scientific forms of knowledge and the "disinterested" inquiry into infinite possible worlds of knowledge that constitute the humanist tradition of the liberal education, on the other. The "state" of knowledge—where the "balance of power" lay between conflicting and competing knowledges—has in many respects reflected the particular society of the historical period under consideration. In broad terms, however, the "state" of knowledge, at least since the establishment of the medieval universities and its claims to autonomy, has been a tension between functional and critical forms of knowledge—or as I have chosen to categorize them—between *instrumental* and *critical* forms of knowledge. These forms have interacted, intermingled and coexisted to varying degrees, as both were seen to have value in their respective domains and both constituted a "good" for society. Instrumentalized, goal-oriented knowledge could be harnessed to create the material and techno-scientific development of society, whereas reflexive forms of knowledge production could act as a check on this, pointing out risk and raising the moral and ethical dilemmas that may accompany the construction of such a predominantly technicized society. Reflexive and critical forms of thought also feed into and inform the social, political and cultural dynamics of the wider world. This critical reasoning is able to not only question and add to other forms of knowledge but is also able to interrogate itself and the ontological foundations of its knowledge production. Indeed this self-reflexivity is its most important function.

In a landmark and prescient volume, *The Postmodern Condition: A Report on Knowledge* (1979), Jean-François Lyotard argues that a momentous shift is now underway, one that is set to change the "state" of knowledge into a lop-sided hegemony by instrumental forms of knowledge and thinking. According to Lyotard, the universities and the information technology revolution hold a central responsibility in this. He writes that "that the status of knowledge is altered as societies enter what is known as the post-industrial age and cultures enter what is known as the postmodern age"(p.3). According to Lyotard, the "technological transformations" in computerization, informatics, telematics, information storage in data banks, etc. "can be expected to have a considerable impact on knowledge"(pp.3–4). He goes on to argue that "the miniaturization and commercialization of machines is already changing the way in which learning is acquired, classified, made available, and exploited"(p.4). And anticipating the arrival of the Enterprise University and commercialization of the learning institutions, Lyotard notes that the old notion that knowledge and pedagogy are inextricably linked has been replaced by a new view of knowledge as a *commodity*, and as a consequence, teaching and learning has become an alienated and alienating process.

Knowledge is produced in order to be sold, it is and will be consumed in order to be valorized in a new process of production: in both cases, the goal is exchange. Knowledge ceases to be an end in itself; it loses its "use-value"(pp.4–5). Lyotard goes on to argue that:

> Knowledge in the form of an information commodity indispensable to productive power is already, and will continue to be, a major—perhaps the major—stake in the worldwide competition for power. It is conceivable that the nation-states will one day fight for control of information, just as they battled in the past for control over territory, and afterwards for control of access to and exploitation of raw materials and cheap labor (p.5).

This commercialization of knowledge reaches into the heart of the university and Lyotard predicts a shift in the whole system of organized learning:

> It is not hard to visualize learning circulating along the same lines as money, instead of for its "educational" value or political (administrative, diplomatic, military) importance; the pertinent distinction would no longer be between knowledge and ignorance, but rather, as is the case with money, between "payment knowledge" and "investment knowledge"—in other words, between units of knowledge exchanged in a daily maintenance framework (the reconstitution of the workforce, "survival") versus funds of knowledge dedicated to optimizing the performance of a project (p.6).

The globalization/ICT revolution nexus has been central to this change. Central, too, has been the concomitant shift from chronologic to chronoscopic time. The temporal shift has geared sense making to forms of knowledge (information) that are rapidly validated (legitimized) and are able to operate on real-time. Lyotard describes "investment knowledge," but based on our criteria of "getting to knowledge," this is more akin to information or, at the very most, surface knowledge. Stock market information can be validated in real-time from others in the field who share the same sort of information, so it becomes information which validates itself, with no real links to the real world, producing its own "knowledge" of the world (from a commoditized perspective) and its own reinforcing logic. The dynamic processes of "information validating itself" to become legitimate "knowledge" are spelling the obsolescence of a capacity for objectivity and criticality; these forms of knowledge, because they are not valorized, cannot validate themselves and thus become "illegitimate," irrelevant, lost, forgotten.

In terms of the state of knowledge production in a commercialized and informationized higher education sector, which then feeds out into the rest of society, Barnett (1997:5) puts the argument succinctly:

> Critical thought cannot be construed just as a form of individual action or mental state. We live in a knowledge society. This *is* the case, but more to the point are the changing forms of knowledge. Humanities give way to science; small-scale forms of knowledge production give way to large-scale forms; knowledge for its own sake gives way to applied knowledge; pure inquiry gives way to problem-solving *in situ*; prepositional knowledge gives way to or at least is supplanted by experiential knowing; and ways of knowing give way to sheer information.

I want now to look at the nature of this "new knowledge production," to try to gauge what it is we may be losing in the shift to a realm of "sheer information," where information validates and legitimizes itself and disregards and de-legitimizes those forms of knowledge that are not immediately valorizable or commodifiable to be placed in the marketplace for sale.

Chapter Nine

The New Knowledge Production: Instrumental Versus Critical Thought

To summarize thus far. We have seen how the interrelated dynamics of the information technology revolution and neoliberal globalization has created the basis for the construction of a new social, political, cultural and technological environment. I have termed this the "information ecology," and it is an environment that has become thoroughly commodified and linked to the logic and imperatives of the market. This environment has its own temporality, the chronoscopic metering of real-time duration, and through the increasing density of interconnectivity is becoming the temporality that we are synchronizing ever more realms of our lives to. The university, as we have also seen, is a central plank in the "information ecology," a crucial node in the global network, so heavily is it commercialized and informationized. And the university, moreover, is a central plank in the production and dissemination of knowledge—old and new, radical and conservative, abstract and concrete, applied and theoretical. But we have seen, too, how much the university has reoriented itself, was *obliged* to reorient itself, away from the project of the place where a "liberal education" could be found, to become a forcing house for industry. Students themselves, in this new time-pressured environment, tend to view university as nothing special, somewhere they go (and pay to go) to acquire a piece of paper that they can use in the job marketplace, the ticket to a job. All these factors, the hegemony of neoliberalism, the suffusion of information technologies, the shift to a chronoscopic temporality, and the wider changes in culture and society that have altered the very function of the university—and the *weltanschauung* of those who pass through them—have changed the processes of knowledge production. Instrumentalized forms of knowledge production, stemming from what Herbert Marcuse in his *One Dimensional Man* termed "technical rationality" are colonizing and hegemonizing those places, spaces and temporalities where reflexive and critical thought and the knowledges these produce used to live. Not just in the universities but also in the media, the culture industries and the public arenas for debate that the university trained usually inform, such as the business sector, the political parties, and the "intelligentsia" as a whole. These are being colonized more and more deeply by the logic that creates and sustains instrumental knowledge production. This generalized effect, the colonization of instrumental knowledge production into the wider realms of politics, culture and society, will be the focus of the final part of this book. I want to preface that, however, by unpacking the terms "critical thought" and "instrumental thought," to get more of an appreciation of what I mean by them and their central importance to us and the societies we create.

Getting Our Bearings Through Critical Thought

In his discussion on critical thought and its role in society, Ronald Barnett (1997:1) uses a powerfully symbolic metaphor to illustrate his point. It is the iconic picture of the lone student standing in front of a column of tanks on the approaches to Beijing's Tianenmen Square in June 1989. "The photograph," he writes:

> depicts a form of critical action on the part of the student. In taking the action, the student is being critical, is entering into a state of critical being. The action is not blind behaviour, but is informed by a knowledge of the political structure of contemporary China and is imbued with a deep understanding of concepts such as democracy and freedom. The critical action is underpinned by critical thought. But, in addition to critical action and critical thought coming into play, the student is purposively taking a stance against the world. The student is saying, literally: "Here I stand. This is the authentic me. This is what I believe." The student has reached this position of brave authenticity through a process of ever-increasing self-reflection, such that he has reached a position of utter assuredness of his own values. The student would have undergone a process of critical self-reflection and has now become fully a critical person.

As Barnett implies in the above, there are three differing dimensions to criticality, to becoming a critical person. These are: *critical action, critical self-reflection* and *critical reason*.

Critical action, as Barnett's representation of the Beijing student implies, refers to being in the physical world, trying to engage with it and working actively to change it. This could be through a wide diversity of functions: from standing in front of a tank to writing a critique of local government parking zone policy and posting it to an Internet website.

Critical action is one expression of the critical person. Another is *critical self-reflection*, whereby through self-analysis and self-criticism one gains the capacity and preparedness to *rethink* one's beliefs, values and perspectives on the basis of new information, new knowledge or the power of new logics. Critical self-reflection brings forth a constant dynamism to the process and stems from the final dimension of Barnett's trilogy, *critical reason*. This is the formal basis upon which the others stand, the "rules and values or theories" (1997:22) that underpin criticality, and the framework that has been built up through the long history of the philosophies of logic and reason from Plato onwards. The university, for almost all its history, was the central institution where the "rules, values and theories" upon which critical action and critical self-reflection depends were developed, inculcated and disseminated. It is no coincidence, therefore, that the Tianenmen Square demonstrations and a whole tradition of Chinese protest came from students and teachers. And not just in China. Students and the university educated have been fundamental in the great social, political and cultural upheavals since at least the French Revolution. This is what universities do (or did) to people. Through the internalization of the dimension of critical action, self-reflection and reason, the individual and the group, through constant reflective dialogue, could "enter into a state of critical being" from where the world could be critiqued and action taken to change it.

The Hegemony of "Instrumental Rationality": Losing Our Way in the World

What is "instrumental rationality"? Quite literally, as the term suggests, it is a mode of thinking, of logic (rationality) that rigidly promotes means-end and goal-oriented actions and outcomes above all others. In the early twentieth century Max Weber developed the term (in German *Zweckrationalität*) as a sociological category and saw it as a central function in the development of the bureaucracy. More recently, as Ronald Barnett (1997:91) observes:

> Instrumentalism takes the world largely as given and attempts to find means of living ever more productively and efficiently in it...instrumentalism works within a horizon of ontological assumptions. The world is objectified: the task is that of securing effects in it and on it. Objects, events, situations, technologies, knowledges and persons are valued so long as they have a use value.

This is a form of rationality, or thinking, that is single-minded and dismisses the possibility or validity of alternatives, of ambiguities or undecidabilities. It is action-oriented and concerned primarily with "productively and efficiently" exploiting the natural world through science and technology and what Max Weber called "rational calculation." In its pure form it admits to no sense of the world being socially constructed and thus, potentially, amenable to reconstruction in another way. Truth, values and knowledge are absolutes, and unchanging through time and space. Having said all this, instrumentalized reason has its place, though—a crucial place. In a world of pure critical thought, nothing would ever get done. Instrumental thought was the key dynamic behind the rise and growth of modernity. But instrumental reason in the project of modernity has always been complemented, to a greater or lesser degree, by critical reason, keeping instrumentalism in check, pushing it off its logical trajectory, that is, toward Marcuse's "technical rationality" and the construction of a "totally administered society" which had its ultimate expression in the death factories at Auschwitz as Adorno and Horkheimer imagined in their *Dialectic of Enlightenment* (1944/1986).

In terms of the hegemony of instrumental thought, it is a question that comes down to, as I alluded to previously, the "balance of power" and where this lies in society at any given period in history. Thinkers from the so-called "Frankfurt School" such as Adorno, Horkheimer and Marcuse were certainly correct in their appreciation of where instrumental reason, on its own, would take humanity. Auschwitz was indeed the end-point logic of a certain kind of rational calculation and instrumental reasoning, with violence and aggression as its dominating features. So, too, could be described Stalin's Soviet Union in the 1920s and 1930s or Mao's Cultural Revolution in the 1960s or Pol Pot's Cambodia in the 1970s. However, these were not instances of the protracted eclipse of critical reason *in toto*. These were isolated and concentrated instances. The instrumental practices of Hitler, Stalin, Mao and Pol Pot could be critiqued, both from within, through underground resistance, *samizdat* literature and open resistance in the streets or from without, through the rest of the world standing against this stripped-down rationality, through sanctions, war or intellectual critique and the leveling of moral opprobrium against the perpetrators and the system-logic they had created and allowed to run free.

These were violent and localized eclipses of "dialectical reason," not an insidious, seemingly progressive and universal eclipse through ICT networks and the domination of the market. This is what is taking place now. Marcuse and Adorno thought and wrote until fairly recently, until the late 1960s and early 1970s, and therefore just before the ICT revolution and the present phase of neoliberal globalization began. And in many respects their analyses were over-determined and overly pessimistic. There still existed in the universities and in wider society deep and radical critiques regarding the nature of society. Indeed Marcuse and Adorno themselves were excoriated by some students, Adorno especially, for not being radical enough, for being conservative and for propping up the old order. Their intellectual demise came quickly after their deaths, and many of their one-time radical detractors fell into the trap of what Guy Debord called *récupération* or incorporation into a system of colonizing commodification that was to spread rapidly through neoliberal globalization, the ICT revolution and the emergence of the Enterprise University. The basis of their thinking was correct—it was just too early and too pessimistic.

Pessimism in theory as in life is a position of powerlessness, an abdication of intellectual duty, and is a position I have not sought to forward here, notwithstanding the negativity of the political, economic and technological dynamics that I have described. My lead in this comes from Raymond Williams (1979:252), who uses critical reason and a deep knowledge of the subject of domination in society to be able to argue that:

> however dominant a social system may be, the very meaning of its domination involves a limitation of selection of the activities it covers, so that by definition it cannot exhaust all social experience, which therefore always potentially contains space for alternative acts and alternative intentions which are not yet articulated as a social institution or even project.

Alternatives—ontological, political, economic and cultural alternatives to the effects of instrumentalized thinking—will be the primary themes in the final chapters. Before coming to these, however, I want to discuss what I see as the effects of a generalized level of "instrumental thought" in society.

Part Three

The Chronoscopic Imagination: Critical Thought and Civil Society in the Twenty-First Century

Chapter Ten

Abbreviated Thinking

In the end it all boils down to a lack of time. Being increasingly suspended in the real-time of chronoscopic temporality means a lack of time to read, to study, to reflect, to consider, to concentrate, to debate and discuss, to care, to empathize, to analyze, to interpret, to scrutinize and to sympathize—and more. We try to do all or some of these, of course, but never properly, never to our own satisfaction or to the satisfaction of others. The result is an emerging real-time culture, where it is increasingly difficult to connect properly with friends or colleagues, where organizations (cultural, business, social) and their constituent members talk past one another, not properly understanding the others' position or needs or aspirations. We supposedly live in an age of ICT "solutions," and what appear as "solutions" to lack of time and lack of proper connection, that is, the "fast-track" online MBA, a faster processor in our laptop, a second cell-phone, a more sophisticated pager or more powerful PDA, Palm Top, BlackBerry or whatever, simply exacerbate the problem, stretching us ever more tautly over time and space, and leaving even less time for us to devote to something, anything, in a more full and complete way. This compulsion toward connectivity, something that in earlier times may have been classified as a form of neurosis, has now been deemed in almost every society an unalloyed good. Indeed, ownership of or access to a networkable device of some kind is viewed almost as a human right. As Andrew Ross (1991:35) observes:

> If contact is possible, then it *must* be made, the circuit of communication *must* be completed—if it can be done, it ought to be done. Why? Not only because people have a collective hunger for the sense of community, but also because in our information society the technical *fact* of communication itself is celebrated as an inherent good. (emphases in original)

However, as Ross implies, the innate "hunger for the sense of community" does not dovetail so easily with this vast and unprecedented technological determinism. In this techno-logic, the increasingly dense and comprehensive networks that constitute the information ecology do indeed connect peoples, cultures, organizations and societies, but, paradoxically, the very act of connecting in real-time also alienates, psychically disconnects and fragments to an ever increasingly fine-grained degree. And this technological determinism shifts easily over into technological reductionism, whereby everything is reduced to the operating level of ICTs and how these can provide the "solution" to the problem, be it the design of a new Nike shoe, large-scale surveillance of a population or the "efficient" delivery of a university education. The technological, the informational and their progeny, the network, become our standard reference points in the information ecology. The danger is that the information ecology that the network constructs is becoming the global non-space, the vacuum of all vacuums, the place of "the final alienation" wherein so deeply have we internalized its logic, we do not even know that we are alienated any more (Eagleton, 1991: 47).

Humans have always interacted with their immediate environment and within the larger ecology that surrounded them. And as we have seen through the work of Ong (1967), this dialectic influenced the development of the "sensorium," the sense-ratio that corresponded, through natural selection, to the demands imposed by the environment in which humans found themselves. Over many thousands of years, variegated ecologies across the planet produced differentiated cultures, cosmologies, religions, worldviews, knowledges and patterns of thought. These physical and social patternings created and sustained the narratives (stories) that are essential to human understanding of the world and of their place in it.

The importance of narrative in our life should not be underestimated. Narratives orient us in the world and anchor us in a form of grounded reality that has its own internal logic that gives coherence to our life and our place in the world. Many fiction writers have understood this, and it has been used to great effect in the works of, say, James Joyce or Marcel Proust. It has been well understood in social theory and philosophy, too. In his 1984 book *After Virtue*, philosopher Alasdair MacIntyre eloquently describes the centrality of narrative in almost every aspect of our lives. He writes:

> We enter human society...with one or more imputed characters—roles into which we have been drafted—and we learn to learn what they are in order to understand how others respond to us and how our responses to them are apt to be construed. It is through hearing stories...that children learn or mislearn both what a child and what a parent is, what the cast of characters may be in the drama into which they have been born and what the ways of the world are. Deprive children of stories and you leave them unscripted, anxious stutterers in their actions and in their words (p.90).

Elsewhere in *After Virtue* MacIntyre quotes with approval the words of Barbara Hardy who remarked that "we dream in narrative, day-dream in narrative, remember, anticipate, hope, despair, believe, doubt, plan, revise, criticize, construct, gossip, learn, hate and love by narrative" (p.173). Narratives are sedimentary; they build up over time, draw upon history, culture, memory and knowledge—all the things in our lives that are laden heavily with temporality and the very things that cannot be sustained in the real-time information ecology. We learn through narratives. The lecturer in the university essentially tells the students a story, or series of stories, about the subject he or she is attempting to explain. They will use metaphors, pictures, signs and symbols that help illustrate that story, link it to elements of our "prior knowledge" and, one hopes, give it purchase and incorporation into our building store of knowledge. Temporality suffuses the whole process, and the capacity for critical thought emerges out from it. Knowledge creates *expertise*.

Here I am not speaking simply of narrow specialization in an aspect of, say, computer programming or tax accountancy procedures but of something more holistic. What I mean is an interaction of knowledge and critical thought that creates dialectical relationships with the environment that surrounds us and the technologies that we create within it. To illustrate this I can do no better than quote at some length the photographer Peter Gullers (cited in Rochlin, 1997:67–68), who writes on the subject of expert judgment of light in photography:

> When faced with a concrete situation that I have to assess, I observe a number of different factors that affect the quality of light and thus the results of my photography. Is it summer or winter, is it morning or evening? Is the sun breaking through a screen of cloud or am I in semi-shadow under a leafy tree? Are the parts of the subject in deep shadow and the rest in bright sunlight.... In the same way I gather impressions from other situations and other environments. In a new situation, I recall similar situations and environments that I have encountered earlier. They act as comparisons and as association material and my previous perceptions, mistakes and experiences provide the basis for my judgment.

> It is not only the memories of the actual practice of photography that play a part. The hours spent in the darkroom developing the film, my curiosity about the results, the arduous work of re-creating the reality and graphic worlds of the picture are also among my memories.... All of the memories and experiences that are stored away over the years only partly penetrate my consciousness when I make a judgment on the light conditions. The thumb and index finger of my right hand turn the camera's exposure knob to a setting that "feels right" while my left hand adjusts the filter ring. This process is almost automatic.

This process, the process of knowledge, expertise and critical thought in action, applies to a multitude of activities in life from fly-fishing, lecturing at university, watching a soccer match, to reading the 5,000-word essay on the international law of the sea I mentioned earlier. For the actor this is reality-grounding, a realm of practice where "values" and "meaning" are given the presumption of permanence or of eternal truths. The practice equates to narratives being recalled and being retold, reperformed and revalidated. In short it is unalienated activity, activity and practice that life in the information ecology, the Enterprise University and the economy of globalization finds fewer and fewer spaces or time for. Our lived experience within the information ecology of real-time duration is obliterating the temporalities required to develop critical thought. We are still human, though, and we still innately respond to and interact with our environment. But the constant "buzz of the flickering present" that constitutes our highly mediated existence within the information ecology is producing new ways of thinking and of being, and as I argued previously, these are instrumentalized and goal oriented in the extreme, nullifying anything that requires time and reflection.

Our interaction with this environment tends toward a process of fragmentation—just as we, paradoxically, become more and more interconnected. For good or ill, the information ecology produces fragmentations of identity, of community, of shared values, of political perspectives and of the narratives through which cultures, peoples, classes, families, even, made sense of their lives in a collective way. From the perspective of postmodern, post-Fordist theory, however, what we are seeing is not fragmentation but an explosion of "diversity" or "subjective autonomy." In other words, a newfound "freedom" from what Lyotard (1979) called the "totalitarianism" of the Western metanarrative concepts of truth, reason and, indeed, rationality itself. It is not my intention to defend the Western metanarrative at the expense of other ways of being and seeing. However, the postmodern perspective ignores the economic realities of globalization and the information technology revolution—and these are profoundly Western and totalitarian in their essence. The commodificationary processes of globalization, powered and enabled by the information technology revolution, have colonized cultures and societies across the world to an unprecedented degree. The market and the logic of the market reach into almost every facet of our lives—especially in the developed economies. What this has meant is that former identities, former subjectivities *have* been broken down (the postmodernist cause for celebration), but they are also being *reconstituted* along the lines of the new hypercommodified order that neoliberal globalization has inaugurated. So deeply has the market penetrated that thinking—thought itself—has become commodified. As Rifkin (2000:55) muses:

> When human thought becomes such an important commodity, what happens to ideas that, while important, may not be commercially attractive? Is there any room left for noncommercial views, opinions, notions, and concepts in a civilization where people rely increasingly on the commercial sphere for ideas by which to live their lives?

Our new subjectivity has been designated "consumer." As consumers the economic forces of globalization relentlessly constitute and reconstitute us as "focus groups," "niche markets," "demographics," "zones," "customer profiles" and so on to be classified, bought and sold just like any other commodity. In the real-time information economy we become database-constructed identities that are continually broken down and reformed in tandem with market-competition imperatives.

There is a further paradox. Whilst the neoliberal globalization-ICT revolution nexus fragments as it connects, it does so within a bounded and homogeneous logic, that of capitalist accumulation and the logic of the interconnected market that pervades much of culture and society. In so doing it creates its own culture: *a culture of capitalism*. In this evolving real-time culture we are losing a sense of sense of our history, existing instead in nanosecond instances of almost pure alienation where we cannot any longer "draw back from our immersion in the here and now" as Fredric Jameson (1995:284) has put it. Immersed in the here and now, in real-time, we have no time to develop knowledge, expertise and critical thought. Without these faculties, the faculties that enable us to live through consciousness in the past and allow us to project possible futures, we are destined to live lives of abbreviated thought and one dimensional imagination.

Abbreviated Culture and Society

The evidences of our abbreviated culture and society are not difficult to find. They surround us every day, so much so that we barely register the kinds of things that only a decade or so ago would have seemed strange, worrisome, unthinkable or simply bizarre. This is not just about our relationship with new technologies; humans have always come to grips with new technologies in one way or another, for good or for ill. It is about the impact of the growing real-time ecology and everything that connects to and from the network; it is about living and learning in this comprehensive and chronoscopic environment—and this is unprecedented; it is about the daily practices—constitutive and constituting of our culture and society—that take place within this electronic ever-present.

Think about the informationized and commercialized university once more. Universities today are strangely quiet places. Wander around them. Individuals sit facing computer screens in libraries, in offices, in hallways, in lecture theatres, in laboratories, even in the parks and lawns in the increasing numbers of institutions that have installed mobile wireless points—almost every nook and cranny contains someone near or at a personal computer. Screens glow and flicker. Fingers race over keyboards that emit their distinctive click. To the uninitiated, these scenes may convey tranquil and industrious intellectual labor, the "getting to knowledge" that universities are supposed to be about. The reality for many engaged with the screen may be somewhat different. They are connected to the network and are working in real-time. Reading and sending emails in abbreviated text, looking for information, writing code, writing and reading reports, reading news, looking for sports results, participating in chat room discussions, almost anything you can think of, for extended periods of the day. But the activity is a form of information skimming, as the term "surfing" connotes. Psychological research into human-computer interaction suggests that we are only able to perceive what we concentrate upon, and when we do not have time to concentrate on a particular thing we are in danger of suffering from Arien Rock and Irvine Mack term "inattentional blindness" (1998). In an information ecology based on real-time chronoscopic temporality, this poses serious problems. Moreover, it is difficult to concentrate on a page of text (scrolling and reading) for more than a few minutes, so people jump from site to site, from application to application, from task to task. Millions of people spend all day doing this, at universities and in regular jobs. As I have argued previously, computerization of the universities, online learning, e-education, whatever the universities have done by way of ICT initiatives has been done as a response to funding cuts and their forced reorganization along business lines. The quiet hum and flickering screens that are the backdrop for most universities today have less to do with learning and a lot to do with competition and aligning with industry.

What is indisputable is that the network gives the universities massive amounts of information. It bulges with hundreds of billions of bytes of digital information flowing at lightning speed from everywhere and from nowhere. How much of this information washing around the network gets beyond the stage of "information," where it can be processed into "knowledge" inside someone's head or onto paper or disseminated as information for others to process themselves? The answer, increasingly, is that if information is not readily graspable, something quickly understood and able to be acted upon, something that can be swiftly utilized to fit the context—we discard it and search, rapidly, for something that is or can be.

As Mads Haahr (2001) has written, "We are less patient, less willing to take the time, perhaps because we have less of it since there is so much we feel we need to digest in order to keep up." But to keep up with what? Haahr goes on to note that:

> We often fail to realize that our interaction with the world is a feedback loop: a circle we can choose to make either benevolent or vicious. As participants in an active culture, we take and we give—this is the core of our interaction with the surroundings. This dual flow of action is everywhere: in language (hear/say), in technology (sensors/feedback), economics (demand/supply), biology (stimulus/response) and computers (input/output). Our current cultural patterns encourage an accelerated mode of interaction: one that expects the rate not only to be high but also to grow. We teach ourselves that speed is good, that a fast-paced lifestyle (busi-ness) is a sure sign of success and that if we can run/work/create faster than our peers, we will do better than them. But acceleration is a risky characteristic on which to base a culture, because a continually tightening feedback loop will eventually become too tight to work well. For the feedback loop to work at a human level, we need time to reflect and digest; to distill information into knowledge; to turn experiences into experience.

The negative feedback loop that drives our "current cultural patterns" and "accelerated mode of interaction" with ICTs is the root cause of our increasingly abbreviated patterns of thinking. And one can only conclude, as does Haahr, that indeed "acceleration is a risky characteristic on which to base a culture." The more we yield the underpinning of our cultural patternings over to technology (acceleration for its own sake) or seeing technology as "both the means and end of human creativity" as Postman said (1993:61), then the more our "interaction" as Haahr terms it, or "dialectic" as I have called it, is going to be a fast-flowing, one-way street where the partial autonomy implied in the dialectic gives way to a determinism on the part of the technology. This deterministic logic need not be something permanent or something that humans cannot regain some form of control over, but the current nexus between the information technology revolution and neoliberal globalization has certainly inserted the logic of determinism into our relationship with ICTs. Abbreviated thinking and living in a constant present, without the time to "distill information into knowledge," is having its effects upon our individual subjectivity and upon our culture and society more generally. One effect is to compensate for the most damaging elements of abbreviated thinking by valorizing it, by giving it the cultural status of a virtue, a talent and a skill. It was not always thus. Abbreviated thinking, or "quick thinking," i.e., the process of thinking and acting upon the conclusion in real-time, used to connote glibness and superficiality in a person. This wariness of abbreviated thinking goes back at least to Plato who wrote in his *The Republic* that "quick intelligence, memory, sagacity, cleverness, and similar qualities, do not often grow together." More recently, as Nicholas Lehmann (1998) wrote:

> It used to be that the business world was neutral, or even hostile, toward this quality. Especially at big corporations, quickness marked someone as peculiar, high-strung, and unreliable. But now the situation has changed. At least in high tech and finance, quick-wittedness rules [and] this is not a state of affairs that would, in most times and places in history, have been considered normal and healthy.

But today it is. Major corporations such as Microsoft have developed questionnaires that are designed specifically to highlight this quality in prospective employees. Being able to "think quick" has its obvious advantages in the sectors such as high tech and finance as Lehmann notes. But in the accelerated New Economy where "grasping the moment" is everything, such thinking is now the *sine qua non* in almost all sectors of the modern economy. Boiled down to its essence this abbreviated-thinking-in-action is very close to a form of gambling, using snippets of information as one would chips in a casino. If one makes the right choice in, say, the stock market, or in a marketing tactic, or, say, in an investment decision in Canada to open a new university campus in Indonesia, using the market itself as the benchmark for truth or validation, then you may be extremely successful or may get badly burned. One only needs to think back to the mid-1990s and the collapse of the hitherto "venerable" Barings Bank of London to illustrate the latter outcome. Barings was chartered in 1762 and had a reputation for dependability and financial probity—the British queen had her money with them. However, in the age of neoliberal globalization and the information technology revolution, the lure of profit changed the way the company did business, the way it "thought" as an entity. In fact it didn't "think" any more in terms of doing the long, tedious and probably (from the company's point of view) needlessly expensive work of researching its markets and taking a long-term perspective. Instead it allowed one individual, so-called "lone trader" Nick Leeson, total autonomy to oversee its futures market operations in Singapore. A combination of inept (and illegal) practices by Leeson and a lack of risk management strategies by Barings management meant that by February 1995 losses run up by Leeson amounted to more than £800 million, almost the entire assets of the bank. The bank was finally crushed under the weight of its massive losses and was sold to a Dutch bank for £1. Who remembers the lessons of thinking in the short-term today? Who remembers Barings today? Probably not the thousands of small and very large investors who collectively sunk billions of dollars into a plethora of doomed dotcom companies in the latter part of the 1990s, into companies that never earned a cent, many of whom patently never looked like earning a cent, but abbreviated thinking (and good old-fashioned avarice) meant that research on the companies was not done and a long-term and realistic perspective never considered. Many of the "burned" would simply have considered themselves "unlucky" and would not really have understood how or why they lost their money.

Numerous and sometimes spectacular disasters aside, the market itself is still the "test" of reality in so many realms of life, and the value of being "lucky" or the skill of effecting successful short-term decisions in the marketplace cannot be rated too highly. The winners remember only the thrill of winning and the losers get pushed aside and are soon, very soon, forgotten. The losers themselves may not forget losing their shirts, but the lesson remains collectively unlearned in the culture of abbreviated thinking. The system-logic fails to change, ensuring that there will always be another unsuspecting "sucker," the "dope on the phone" as Frank (2001:110) terms them, who can always be relied upon to think, act and open his or her wallet too quickly. The winners will go on to set the measure of success for others to follow and reinforce the logic and consequences of abbreviated thinking. The problem is that it further foreshortens the horizon of the possible and means quite literally that many previously important questions are now "unthinkable." There is less and less time, individually or collectively, to think through many deeply important issues. Questions such as: What realistic alternatives are there to the "free-market economy"? Is it any longer possible to have a society that is not capitalistic? Do we have to informationize, wall-to-wall and floor-to-ceiling, when we don't really understand the consequences? And so on. These are only some economic questions that come to mind. There are many other cultural, ethical, political, moral and spiritual questions that are no longer on the agenda because they are fundamentally "non-commercial" and so are outside the realms of thinkability.

Chapter Eleven

"I've No Time for Politics!"

The well-worn but little reflected-upon cliché that "politics is the art of the possible" has begun to take on new and troubling dimensions in the truncated political imagination that stems from living and learning in the information ecology. In short, the scope of the possible has narrowed in the minds of many. For increasing numbers of individuals, young people especially, politics has become anathema, something practiced by corrupted or glib men in suits, in it only for themselves. In the Anglo-American societies especially, in election after election there is a drop in the numbers of people who vote. The 2000 US presidential election attracted the lowest number of votes in over seventy years—less than 50 per cent of eligible voters "elected" George W. Bush. In Britain, the 2001 "landslide" re-election of Tony Blair's New Labour was achieved by a "mandate" from only 25 per cent of those 44 million persons eligible to vote. It was a victory for the "Stay at Home Party," according to the political analyst for the *Daily Mail*, Edward Heathcoat Amory. Moreover, survey after survey continue to indicate public cynicism and/or apathy regarding the political process and the politicians who make it happen. Mainstream party membership has dropped correspondingly, revealing that people don't want to get involved any longer. Why?

It is not my intention here to argue that there were halcyon days in the past where the "political sphere" was vibrant, democracy was being expanded and strengthened, and participation was widespread. The reality, since at least the time of the Industrial Revolution, has been much more complex, and has been one of class struggles, the rise and fall of ideologies and, centrally, the changing social and political dynamics stemming from the dominant mode of production at different historical phases.

We may begin this discussion with a grounding definition—"What is politics?" Like the question "What is time?" that I posed previously, it may seem to be obvious: politics is what they do in the Parliament building or in the Houses of Congress, and, partly, it is. Indeed this may be part of the problem today, the fact that politics is viewed by those many respondents to surveys on attitudes toward politics and politicians with a mixture of cynicism and loathing or complete indifference. People, young people predominantly, view this particular aspect of politics as remote, something that does not speak to them or their concerns. As Robert McDonald and Frank Coffield (1991) write:

> Politics for most young people means a drab world of grey, be-suited, middle-aged, middle-class, male MPs, compulsory party political broadcasts and strange, heated arguments about which they know little.

A good many not-so-young people will have this perception of politics, too. However, let us dig a little bit deeper to attach a working definition to the term and then go on to give this some historical and theoretical purchase. Politics, at its most abstract, generalized and possibly utopian level is for me about individual and collective participation in the public sphere, geared toward the promotion of democracy. What, you will now ask, is *democracy*? Theories of democracy have a long history with a great many radically divergent views that there is neither the space nor vital need to cover here. Again, a working definition must suffice. Democracy today is the hegemonic political form across the world. To make things a little bit more complex for the purposes of my argument, it needs also to be said that democracy as we know it is linked to the historical process of *modernity* where the individual acting as a rational subject evolves to formulate and construct universal and secular first principles. These are the laws, rights and standards that recognize and seek to develop further the liberty, freedom and equality of the individual in society. Politics, evolving in the ever-changing dynamics of modernity, is the *mechanics* through which democratic culture emerges. This ideal however, has rarely, if ever, materialized in its totality. The outcomes, moreover, need not be in any way democratic, as witnessed in the politics of fascism, where in its Nazi variant democracy was viewed as weak, and in Stalinism, where the principles of democracy were distorted, in practice, beyond any recognition.

I want now to briefly trace the development of politics and political engagement in the West since the emergence of the Industrial Revolution. Analysis of this history will show, I believe, how and why political engagement today is of a radically different order than from, say, as recently as the 1950s. I will also show that this change has its roots in the shift from chronologic to chronoscopic time, and the interrelated dynamics of neoliberal globalization and the information technology revolution that gave effect to that momentous temporal shift.

Chronologic Politics

> *The liberating force of modernity is exhausted as modernity triumphs.*
> Alain Touraine, *Critique of Modernity*
> (1995: 91)

Prior to the rise of modernity, the Industrial Revolution and the spread of Enlightenment thinking, politics was a rather elite game in terms of the number of those able to participate. It had little to do with voting or democracy and a lot to do with attaining power and wealth—and politicking to keep it. In feudal times, politics in its "low" form was constituted around the petty struggles between non-peasants over such things as land, title and wealth; in its "high" form it revolved around diplomacy and interstate relations in the "great game" such as described in Niccolo Machiavelli's *The Prince*. With the rise of modernity, these struggles became wider, the cast of actors more inclusive and the stakes higher. In Britain initially, improvements in communications technologies began to spread ideas, materials and wealth throughout society. These processes contributed to the evolution of an economy that was expanding, becoming more organized and yet more integrated. It was from this basis that a nascent capitalist mode of production began to replace the thousand-year-old feudal order. As modernizing and industrializing societies grew, they became more integrated. And as this process developed, it became increasingly difficult to escape the effects of these world-changing dynamics. Peasants became workers and the isolated became collectivized. Ordinary people, through being part of a formative class with an increasing power potential, became a political force for the first time in history.

From the middle of the nineteenth century onward, the Industrial Revolution got into full swing. The needs of the growing capitalist system and the struggles of the competing actors within it led to a range of profoundly important social and cultural developments. Education began to spread through society as it was realized that more people being able to read and write, as well as having a degree of technical literacy, was vital to grow and develop the economy still more. With the rise in literacy came the rise in mass media, in newspapers, magazines and so on that people could use to develop relatively complex ideas on the nature of society, its sciences and technologies—and its political processes. Democracy was increasingly on the agenda in the industrializing societies of Europe and North America. With the collectivization of the worker came a concentration of social power, one that was able to challenge the rising capitalist class. These were the dynamics underpinning the organized expression of this class power, the rise of mass political parties and the potential for collective political engagement for those former peasants who would have been excluded from the process altogether. "The people," through their political parties and other organizations, began to press for more inclusivity, more of a share of the growing wealth, for an extension of the franchise and so on. As more and more educated, skilled and literate individuals comprised the growing modernizing and industrializing society, then so too did the politics of the society become more diverse and complex. The political imagination, in other words, was constantly widening in its scope, and the realm of "the possible" was felt in many quarters to be limitless. From the middle of the nineteenth century, the "liberating force of modernity," as Touraine called it, was arguably at its zenith. Part of its power may be described in terms of a "discursive flux." Michel Foucault, in his *Archaeology of Knowledge* (1972), argued that the structure, the set of rules concerning what could legitimately be discussed regarding *knowledge* in a society is established by particular "discursive practices." Within modernity, the dominant "discursive formation" was centered on that of the immutability of particular ideas of progress, reason and liberal democracy as constituting a linear path toward social perfection. However, within this hegemonic discursive formation was a range of differing bodies of knowledge, differing interpretations of reality that competed for legitimation as the only "truth" and "reality."

From the mid- to late-nineteenth century, a flowering of "discursive practices" that were expressed as political and philosophical ideologies and movements occurred. This constituted a historically unparalleled phenomenon. During this time, capitalism was by no means secure as the dominant social mode of production and hegemonic discursive formation. It was prone to violent swings between boom and depression, and so reforms or wholesale alternatives were constantly being sought and propounded. Critical thought flourished, and the period saw the emergence of various political ideologies and movements that took hold to a greater or lesser degree in Europe, the Americas and elsewhere. Movements such as anarchism, syndicalism and socialism in its democratic, revolutionary, Fabian and Christian varieties evolved and made their impact upon society, with the echoes of many of them still with us today. At the same time, the rationalizing force of modernity organized workers into unions, pressing for economic demands. Unionized workers and intellectuals soon took the next logical step and organized themselves into political parties. The "masses" were organizing and struggling in the political realm in response to the depredations of industrial capitalism, and so much of the motive force of politics came from the workers and the social democratic traditions they were establishing. The first mass organized party along these lines was the German Social Democratic Workers Party, established in 1869. In Britain the Fabian Society began in 1883, the Independent Labor Party in 1893, and the Labor Party in 1906. In the USA, the Socialist Labor Party of America was set up in Philadelphia in 1876. Across the industrial world, in proportional numerical terms, the membership of mass political parties (social democratic, revolutionary, anarchist and so on) was at its height around the end of the nineteenth century. Eligibility to vote was becoming less restrictive, bringing more men into the political process. Long-standing political struggles by suffrage groups in the advanced economies eventually bore fruit and the franchise was extended to women, bringing them into what was by now the mainstream of political activity. Women won the right to vote in the USA in 1920. In Britain women over the age of 30 were given the right to vote in 1918, and in 1928, after mounting pressure from women's groups, this was changed so that all women had equal political rights with men.

These and many other forms of activity constituted politics at its most historically dynamic and dramatic and with the "liberating force of modernity" at its most potent. Clock time was the organizing principle of this phase of modernity, and clock time was eclipsing other temporalities to become the predominating time-sense of most people, organizations and institutions. There was a feeling of the "march" of time, of progress and of a future that contained the prospect of better times. Political activity was contributing to this dynamic and was occasioning momentous changes in many aspects of culture, the economy and society. It could be argued, indeed, that the one hundred years from around 1850 until 1950 constituted a century of modernity at its most vibrant. Marx (Marx-Engels, 1975:38) captured the *zeitgeist* most famously when he wrote that:

> Constant revolutionizing of production, uninterrupted disturbance of all social relations, everlasting uncertainty and agitation, distinguish the bourgeois epoch from all earlier times. All fixed, fast-frozen relationships, with their train of venerable ideas and opinions, are swept away; all new-formed ones become obsolete before they can ossify. All that is solid melts into air; all that is holy is profaned....

In the twentieth century this momentum was punctuated by two world wars that shattered the existing political landscapes and reorganized them into different forms after the conflicts had ended. After the First World War, and especially after the Russian Revolution, socialism and workers' power was again on the agenda as the carnage of the war was in many places attributed directly to capitalist greed. Socialist revolutions broke out in a range of countries in Europe such as Germany and Hungary. Political turmoil became the "norm" for many countries across the world, not least in the leading industrial nations such as Britain, France and the USA as well as those countries still under the dominion of European imperialism such as India and China. Economic crises added spectacularly to this broad political unrest with the stock market crash in the USA in 1929, an event that set in train a series of economic, political and military crises that culminated in another general war in 1939.

The three decades immediately following the Second World War could be described as the period when the liberating force of modernity actually began to exhaust itself. A *discursive shift* began, a shift toward the consolidation and hegemony of the ideology of capitalism and the idea of a liberal democratic society. "Exhaustion" may seem a slightly peculiar noun with which to describe the "golden age" of capitalism, a phase where profits steadily rose, workers were getting better pay and conditions and, materially speaking, people in the West "never had it so good." But in retrospect, this material abundance (and this, principally, was *all* the "golden age" amounted to) masked an increase in the relative poverty in other areas that may be said to contribute toward the more multi-dimensional human existence. Marcuse railed against this contradiction in his *One Dimensional Man* (1968). In it he argued that advances in technology, the material provision of more cars, jobs, money, housing, gadgets etc. had reached such an extent that it had become a form of social control and the means for increasing alienation: "Our society distinguishes itself by conquering the centrifugal social forces with Technology rather than Terror, on the basis of an overwhelming efficiency and an increasing standard of living" (1968:9–10). With basic needs satisfied and an instrumental rationality inserting itself into every realm of life, the analytical and conceptual frameworks needed to explore existential questions, questions of who we are and what we could be, became sublimated, nullified and increasingly forgotten. A one dimensionality began to assert itself, Marcuse argued, as the status quo of the bureaucracy, the political system and the media became more entrenched. Particular interests were increasingly "sold" by the mass media and the political process as "those of all sensible men" (1968:9). Modernity, reason and progress become clichés as bureaucratization, mechanization and routinization become the real motive power in society. The intrinsically *collective* nature of politics increasingly dissipated as individualism was asserting itself as the cultural and social practice. The result was a diminished political sphere, where:

> The political needs of society become individual needs and aspirations, their satisfaction promotes business and the commonweal, and the whole appears to be the very embodiment of Reason. (1968:9)

It was on the left of politics, where "critical thinking" and "critical theory" were the primary frameworks of analysis, that alarm bells were being rung during the 1960s. Guy Debord was one who, like Marcuse, saw modernity as having exhausted its libratory potential and become a process of domination over the population through the commodity form and what he termed "the spectacle" which consists of an endless diet of diversion and trivia, imprisoning the individual in the negation of the original promise of modernity and democracy. In *The Society of the Spectacle* (1967/1983:6) he argued that:

> The spectacle, grasped in its totality, is both the result and the project of the existing mode of production. It is not a supplement to the real world, an additional decoration. It is the heart of the unrealism of the real society.

Modern society was interpreted quite differently by leading social theorists from the conservative wing of politics, such as Daniel Bell. Writing in 1962, Bell, in his book *The End of Ideology*, argued that liberal capitalism had triumphed in the Western world. So much so, that there was a "rough consensus" among intellectuals on the primary political issues that should go to make up the desirable society: "the acceptance of a Welfare state; the desirability of decentralized power; a system of mixed economy and of political pluralism" (pp.402–3). Echoing somewhat Harvey's comments about the psychological and political effects of post-war Fordism, Bell argued further that "The new generation...finds itself seeking new purposes, *within the framework of a society* that has rejected, intellectually speaking, the old apocalyptic and chiliastic visions" of the ideologists of the left such as Marcuse and Debord. Notwithstanding their differing perspectives upon the fate of modernity and its political vitality, all these writers—and many more—implicitly or explicitly acknowledged that society was no longer moving forward, that modernity was no longer progressing toward the perfect society. For Bell and other conservatives, we had already reached something like the perfect society, and the "rough consensus" was but a reflection of our satisfaction with liberal democracy. For Marcuse, Debord and a host of others who came after them, the end of progress meant not that we had reached the uplands of prosperity and the final self-realization of human subjectivity, the optimal blend of personal freedom and material plenty, but a reversal of the modest gains that modernity and mass chronologic politics had achieved since the emergence of the Industrial Revolution and the Age of Enlightenment.

The particular "discursive formation" that viewed capitalism, liberal democracy and the mixed economy as the basis of the only reality was becoming more and more concentrated as the "golden age" of Fordism was reaching its "high Fordism" apogee, and the total "way of life" it was to become during the 1950s and 1960s. Notwithstanding the intellectual influence of the works of critical thinkers in the universities and beyond, they were, in practical terms, wholly marginal to the lives and thoughts of most people. In politics their influence was similarly marginal. Mainstream political thought and action was concerned with empirical problems of the economy and of working for reforms inside the boundaries of the liberal democratic model. Socialism and communism in their varying stripes were at this time either secondary and ineffectual as centers of alternative thought or were carbon copies of the rigid and generally discredited forms of socialism that existed in the USSR or China. Genuine alternative political thought and action such as environmentalism was in its nascent stages of development during the 1950s and 1960s—and earmarked for *récupération* once the ideas gained a certain level of popular acceptance. Critical thinking was being squeezed out of the discursive formation, a formation that was becoming narrower and more self-satisfied.

The End of Chronologic Politics: The Eclipse of the Theoretical by the Practical

Chronologic politics were the politics of modernity, and modernity with its overarching dynamism was ideologically committed to, as well as functionally oriented toward, the construction of the ideal society. This was a universalizing gospel, where the construction of modernity, through politics, was to transform the world, to haul it from backwardness, filth and disease toward a world of light, of steel, of machines and of plenty. Throughout its one-hundred-year phase of "liberation," modernity and its discursive flux of competing ideas and alternative ways of constructing the ideal society was, as Marshall Berman (1982:243) wrote, "modernization as *adventure*." For many of those engaged in the political process as politicians, civil servants and so on, the job, not unlike the priesthood, was viewed as a vocation, and an adventurous one at that, where the reward was to be part of history, to be a part of making the world a better place for everyone. Politics was not only part of the great "adventure" that was building the future, politics also had its *past*, its traditions, its pride, its heroes and its villains, its winners and its losers. Chronologic time was central to this process. It helped to create a narrative, a context to develop an understanding of a history, of a glorious (or sometimes not so glorious) past, of an exciting and meaningful present, and the possibility of a splendid future for all.

Berman also acknowledges that modernity has its other side, its *alter ego*. The flip side of modernity as adventure is "modernity as *routine*" (p.243). Like the "technical rationality" discussed earlier, "routine" has been as indispensable part of modernity and the politics it had spawned. "Routine" kept "adventure" in check, just as instrumental rationality kept pure reason in check. Routine got things done, whereas the adventure constituted part of the abstract ideal, something to be aimed for. However, in the chronologic politics of modernity, routine began to establish itself early on. In most of the modernizing democracies, the party system crystallized around the establishing parties, those who organized soonest and accumulated a social and economic power base with which to crowd out the others. Accordingly, two or three main parties, representing differing power interests within the framework of bourgeois liberal democracy, became the norm. And in most of the social and liberal democracies today (USA, France, Britain, Australia, Germany and so on) the main parties are the direct descendents of those established in the last fifty years of the nineteenth century. The only recent alternative to the years of the nineteenth century. The only recent alternative to the established party system has been the environmentalist "green" parties, whose fortunes have waxed and waned across the Western democracies. Moreover, their rhetoric (if not their policies) is always in danger of *récupération* by the established parties should they become too popular. Periodical elections are another major feature of the routinization process. Three, four or five-year timeframes are now the norm for the electoral cycle when voters must choose another party to govern from these two or three established parties (usually two) with a realistic chance of forming government.

As the interrelated dynamics of Fordism, liberal democracy and capitalism became more entrenched, more complex and more of a "way of life," they also became more deeply routinized. "Adventure" inexorably gave way to "routine," especially during the period of "high Fordism" from 1950 to 1975 when the mode of production was at its most comprehensive. The politics of the period reflected the mode of production and its increasingly sluggish dynamics. What came to be known as "machine politics," aptly reflecting the Fordist culture that helped propagate it, became the institutionalized political culture. Dwindling or extinct was the vocation politician who saw liberal democracy as a project. In his (usually male) place came the *apparatchik* who would toe the party line as opposed to constituent concerns, or the self-interested careerist intent on building his own power base, focusing only on the intrigues and alliances that can help him reach the top. This process did not happen overnight, but like the system of Fordism it reflected, it was one of gradual decay, of systemic maturation and growing crises. Giddens (1994:90) called this process "the inertia of habit," a stifling political lassitude that hung as heavy as rain clouds over the Western democracies during the period of economic boom and "high Fordism."

The discursive flux that constituted the various ways of being and seeing that represented "modernity as adventure" had solidified into a discursive hegemony that was expressed as "modernity as routine." And so it was with politics. What US president George Bush was later to deride as "the vision thing," that is to say, actually having a motivating *dream* and conviction about what the world *could* be, was reduced to tinkering at the edges of a maturing and crises-ridden system. Routinization and technical rationality began to snuff out the discursive flux and the alternative ways of being and seeing they were able to sustain. In politics this was a problem, not only for the political system itself but also for the civil society it helped create. The social world and the knowledge it produced, as Marcuse and others had observed, were becoming one dimensional. Pragmatism and utility increasingly canceled out theory and critical reason, and politics increasingly operated within the realms of its own narrow logic, the logic of liberal democracy. As Pierre Bourdieu (1990:214) observed:

> Knowledge of the social world and, more precisely, the categories which make it possible are the stakes par excellence of the political struggle, a struggle which is inseparably theoretical and practical, over the power of preserving and transforming the social world by preserving or transforming the categories of perception of that social world.

The "inertia of habit" that stemmed from the maturation of Fordism and its hegemonic discursive regime meant that, amongst many other things, the "practical" eclipsed the "theoretical." Chronologic politics was practically oriented and geared toward the preservation of the bourgeois social order as had been made clear by a great many writers from Marx onwards. An unanticipated consequence of this, however, was the preservation of not only the social order but also its "categories of perception." The metaphor of the machine not only fitted the productive regime of Fordism but also the politics and the *thinking* of the time. The alternative politics, the alternative "categories of perception" that would burst into life during the 1980s and 1990s were, for reasons I shall come to, marginal, extremely marginal, during the 1950s and 1960s. Given the scale of the crises of Fordism during its demise in the 1970s, however, it was not possible any longer to preserve the social world, and the "categories of perception" that bounded it were to be forcibly transformed. Once these categories of perception were transformed, Fordism and the way of life that it had sustained for so much of the twentieth century would be gone forever—as would its distinctive politics. A new politics would arise, the composition and future of which are not yet clear. They have their genesis in the crises of Fordism, however, the resolution of which led to the shift from chronologic to chronoscopic time.

Chapter Twelve

Disorganized Capitalism and the Emergence of Chronoscopic Politics

> *The trouble with the zealots of technology as an instrument of democratic liberation is not their understanding of technology but their grasp of democracy.*
> (Benjamin Barber, 1997:224)

In their 1987 book *The End of Organized Capitalism*, Scott Lash and John Urry put forward what was at the time quite a novel thesis. It was one that from the perspective of these "left" thinkers was very much against the grain of "traditional" left-wing thought on the nature of capital and capitalist society. The traditional account of capitalist development, dating from the early twentieth-century analyses of writers such as Rudolf Hilferding and Jürgen Kocka, saw the logic of capital, and by extension capitalist society, as evolving increasingly toward "organization." This organizing mode of production grew out of an earlier, more undiluted and freebooting form of capitalism, where there were very few rules and very little coordination in and between markets, distribution, banking, finance and so on. According to Lash and Urry's thesis, capitalism began to "organize" during the last decades of the nineteenth century. The change, they argue, was in response to the deep economic slump that world capitalism suffered during this time. The process of organization did not occur simultaneously in all Western countries, nor in exactly the same way. But capitalism across all industrializing economies *did* start to exhibit common features of economic and social organization. The most prominent of these were: the concentration and centralization of capital (industrial, banking and commercial) as markets became progressively regulated; a separation of ownership from control of capital; the emergence of bureaucratization and managerial hierarchies; growth of collective organization in the labor market, that is, in trade unions, employers' associations, professional guilds and so on; and various ideological changes concerning the role of technical rationality and the valorization of science and technology (1987:3–4). To this could be added the suffusion of clock time as the standard organizing principle across industrializing and modernizing societies.

This process was well understood and broadly accepted (at least on the left) as the "natural" developmental trajectory of capitalist development. As late as the 1960s and the 1970s this was still the standard "left" analysis: that is, capitalism, and by extension capitalist culture and society, was becoming more organized, more bureaucratized, and more calculable and predictable in its onward trajectory. Indeed in 1966 Paul Baran and Paul Sweezy published their highly influential *Monopoly Capital*, which carried forward, with a contemporary analysis of American capitalism, the tried and trusted thesis of the organization principle as the primary dynamic of capitalist development. Much of this economistic investigation was in many respects a paralleling of the philosophical and sociological analysis of Marcuse's "one dimensionality" and the prison cage of technical rationality, and Harvey's retrospective characterization of Fordism as a whole "way of life" that we have discussed, and is testimony, I believe, to how deeply society had internalized this particular mode of production.

However, Lash and Urry argue that by the mid-1980s this "organizational" logic had run its course, and capitalism was entering a new phase of what they term "disorganization." They describe fourteen main features that constitute the "disorganization" of capitalism, the main ones being: the *de*concentration of capital from the point of view of national markets, and the near-universal decline of cartels; the international expansion of large corporations; the continued rapid rise in the numbers of white-collar workers, and a distinctive service class of managers, professionals, educators, administrators and so on; a decline in class-based politics and a concomitant rise in so-called "social movements" such as environmentalism, anti-nuclear campaigning, women's movements, etc.; a drop in the absolute and relative size of the working class and of manual workers in manufacturing; a shift from "Taylorist" to flexible forms of work organization; and an increase in cultural fragmentation and pluralism and the development of new political/cultural forms (1987:5–7). In essence, although they do not make the point explicitly, what the authors are describing in the shift from "organized" to "disorganized" capitalism is the transition from Fordism to post-Fordism that has been the backdrop to much of this discussion. However, nowhere in the book do they analyze or describe Fordism—the supposed object of "disorganization," a factor, I believe, that prevents them from getting to the real core of the problem, that is: the deep changes within that regime of accumulation. This aside, both the authors and I are at one in respect to the *effects* of "the tremendously significant transformations of time and space, of economy and culture" that have "disrupted and dislocated" the patterns of capitalist development (they call it organized capitalism; I call it Fordism) that dominated for much of the twentieth century (1987:2).

In politics, of course, disorganization was already underway by the 1960s. In retrospect, "the sixties" became a metaphor for rebellion, alternative lifestyles, "dropping out" and so forth. The French have a term for the sixties and the generation that grew up in them; they call it *soixante-huitard*. The year 1968 became famous (or infamous, depending upon your political affiliations) as the high watermark for the counter-cultural and political challenge to the "system." This was the point where the USA in particular was "at the crossroads," and the Western democracies more generally were in social and political tumult. Again, depending upon where you stood on the matter, the late 1960s represented everything that was wrong with the system, what a lack of order, respect and discipline could spiral into—or it represented a tantalizing glimpse of other possible ways of being and seeing. Whatever it was, the social, cultural and political turmoil of the period signaled the beginning of the end of an epoch, the end of a center that could no longer hold, and the collapsing of the centrifugal force of modernity.

The universalistic politics of mass parties, of shared perceptions of the future and a belief in the ideas of progress began to lose its all-encompassing potency in the West. This was not a sudden, cataclysmic paradigm shift toward the embrace of a radical alternative to modernity and all it stood for. It was more like the slow-motion *fragmentation* of the edifice of modernity, with pieces of it arcing off in every direction. As the "way of life" that was Fordism began to crumble during the 1970s and 1980s, new forms of politics, what were later called "new social movements," began to emerge. In line with the fragmentation of the universalistic principles of modernity, these pursued the politics of single issues, such as women's liberation, animal liberation and the various groups concerned with aspects of the environment. Many began with much enthusiasm from many young people who had been disillusioned with the old order and its "totalitarian" metanarratives. Politics became "personal," and so the politics of race, of gender and so on began to eclipse the politics of class and class power. In many ways, then, these micropolitics took the focus off the real class and class power battles that were raging during the 1980s and on into the 1990s—the battle over the reconstruction of capitalism which we have discussed at length. It is no exaggeration to say that if political struggles and organizations had not fragmented into a thousand different single issues, then what turned out as massive defeats for organized labor and working people in general over the reconstruction of Fordist capitalism could have turned out differently. But it gets worse. The rise of new social movements, of single-issue politics, a postmodern worldview and a disdain for anything tinged with "class" or socialism meant a degradation of the intellectual ability to connect single issues to more overarching issues; it meant the degradation of a sense of history and a more holistic understanding of how issues develop, and, crucially, in the context of our discussion, it meant that the generations that began to grow into adulthood during the mid-1990s had little with which to intellectually prepare themselves to critique the neoliberal globalization/ICT revolution nexus and its effect upon politics and society in our own time. Moreover, the effects of this nexus in the universities, with their inculcation of generalized instrumental and abbreviated thinking in students, meant that they would be doubly unprepared for politics in the Age of the Internet.

Techno-Politics: Seattle and All That

The primary questions to consider are: Does the information ecology, with its information-sharing, idea-disseminating, organization-building and paradigm-shifting potential, signal a new phase in the processes of political struggle, where people can be more empowered and where, finally, the massive defeats across the broad social, cultural and political fronts that were inflicted upon ordinary people all over the world since the late 1970s can be reversed? Can these powerful networks of interconnectivity, created primarily for profit, be turned against their creators, the global corporations, and used as vectors of equality, justice and fairness for the majority of the people? Many think that they can, and hundreds of thousands, if not millions, of people are acting upon this belief.

It is claimed that global and globalizing networks of activists are now able to use the accoutrements of info-capitalism such as the cell-phone, pager, email, Internet and so on as powerful tools of resistance against social fragmentation and political totalitarianism. Proponents maintain that the limits of the single-issue politics of the new social movements that emerged in the 1960s and 1970s have finally been recognized and that techno-politics represents the birth of "a unified movement of holistic, social, economic and political change" (cited in Klein, 2000). And, indeed, something different *is* happening. The end of the twentieth century saw the emergence of the so-called "anti-globalization movement," a coalition of pressure groups, unions and political movements of every stripe and from every part of the world. Some commentators saw this development as partly indicative of the emergence of a "global civil society." For example, the number of international non-governmental organizations (INGOs) across the world has grown by nearly 25 per cent since 1990; their membership has grown by 72 per cent and the number of links they have with each other has doubled. Moreover, the number of civil society events that paralleled government summits has grown fourfold, and the numbers of participants in these and other events that focus on global issues has soared (GCC Yearbook, 2001).

This global civil society movement simply could not exist without the Internet and its connectable bits and pieces. The anti-globalization movement and the explosion of INGO political and organizational activity is a direct response to the perceived depredations of economic and cultural globalization. But can it *change* anything, can it have a *real* effect—and most importantly, are techno-politics the answer to the genuine need for "holistic social, economic and political change"?

To try to answer this question we need to think critically about the nature of the intersection of the two principal elements that have spawned this "new politics," that is, globalization and the ICT revolution and their effects upon social and cultural life. What is the very essence of the neoliberal globalization/ICT revolution nexus? The answer, as I have tried to make clear throughout these pages, is the insertion of the powerful logic of *instrumental rationality* and *efficiency* in the accumulation processes of capitalism. This logic has done its work on the productive régime of capital accumulation, the labor process, on the movement of capital, and has inserted itself as the organizing principle in almost every realm of culture and society *as well as* the economy. It has contributed massively, as Gleick has termed it, to the "acceleration of just about everything," and has been the major constitutive factor in the manifestation of the information ecology of real-time duration. Political discourse and organization have not escaped from this process, be it on the margins where new social movements operate or in the realm of the mainstream established parties. As I will show below, this fact has set up some major problems for both social movement and "traditional" forms of politics.

At the beginning of this final part of the book, we discussed what politics is about, its aims and its principal dynamics. Further questions need to be asked, however, in respect of the analysis of politics in general—and techno-politics in particular. What *sustains* politics? What is it that gives it its motivation, its dynamism and its energy? What makes it, potentially, an enormously powerful vector for social, economic and cultural change? The answer is *communication*: be it oral, written, printed, electronic or whichever technologically mediated form that human ingenuity has devised over the last ten thousand years. The myriad ways in which people can communicate, one-to-one, one-to-many, many-to-many, are the essence of politics. Communication generates ideas, ideas can crystallize into meaning, meaning can be the spur for action, and action feeds into the processes of change. And ever since humans developed language and rudimentary forms of writing, technology has been a fundamental determinant of the shape and nature of the communication process within society.

This framework for analysis, that is, one that sees that technological systems have profound effects upon culture and society, may seem a fairly obvious one. However, the analysis (and, hence, the relationship between technology and society) takes on new significance when dealing with politics as constituted through digital communication systems. Human communication has just been subjected to the hyper-instrumentalizing and super-efficiency logic of ICTs, a development unsurpassed, probably, in its significance, since the introduction of the moveable type press by Johannes Gutenberg in the 1450s. ICTs are paring back human communication to its bare minimum—to the transmission of *information* from one communication point to another. This *instrumentalization of political discourse* is an extremely important change. Before this development, the transmission of the discourse of politics, its ideas and their dissemination through various communicative modes had vital elements that could not easily be dispensed with. Primary amongst these were physical presence, space and place. Political discourse at its most elemental level was conducted face to face, as part of a crowd, as a member of an organization that had specific physical connections such as the workplace, the party room or the meeting-place. This physical relationship involved, to some degree, a committed relationship, a social pressure to act in a certain way and to have communal responsibilities toward certain ideas and the people who share them. And, like the processing of data into information and information into knowledge, these things take time. Indeed, relationships of *trust and commitment* (in the Fukuyamaian sense of mutuality) build up through both *temporal* and *spatial* dimensions—the very dimensions of human communication that ICTs have the profoundest effect upon.

Let me explain. We can understand this more clearly, perhaps, if we take it to the level of the everyday. Imagine a new neighbor moves in next door. Of course, you'll smile and say "hello"; you may even (although this is possibly becoming rare and might even be considered forward or rash) invite the new neighbor into your house for a cup of tea. What you would not do is automatically trust that person. You would not, for example, ask them to baby-sit for you that night—it may take years to get to that stage—if ever. Nor would you easily be able to make any sort of commitment to that person at this stage, such as offering to baby-sit for your new neighbor yourself. You wouldn't do it and they wouldn't expect you to—trust and commitment don't exist. The non-existence of trust would not be paranoia; this is just the way societies work, and what is more, most of us understand that getting to the stage of trust and commitment takes time and effort—weeks, months or years—but not in realtime. The spatial dimension is important, too. Relationships of trust and commitment simply do not progress in the same way and at the same time if there is not close spatial proximity. Trust and commitment build up through place and space as well as time. I should make it clear at this point that I am not arguing that trust and commitment are only achievable through people living close to each other over a long period of time. I am arguing only that when these disappear, then trust and commitment are vastly more difficult. For example, making a commitment to someone's face, someone you have known for even a short time, carries a far greater feeling of commitment (or obligation) than saying you'll do something in an email to someone you have never met. How often has someone said by email that they would do something—and then never have them follow up? How many times have you done it to someone else?

Transfer this thinking to politics. Politics based upon temporally formed bonds of trust and encompassing the committed and anchored relationships that emerge from proximate political discourse has been, for at least two hundred years, a powerful agency for change. Indeed, this was central to the forms of class-based politics discussed previously. As Mario Diani (2000:119) has argued:

> Participatory movement organizations—especially the most radical—are more dependent upon direct, face-to-face interactions, both for the purpose of recruiting members and securing their commitment. Engaging in what are potentially high-risk activities requires a level of trust and collective identification unlikely to develop if not supported by face-to-face interaction.

Political discourse based upon the instrumental logic of ICTs winnows out these organic connections that give political action its power. What is left is almost pure communication, with the fundamentally important elements of trust, commitment and political responsibility hollowed out. This is real-time, techno-politics. Certainly, techno-politics can bring people together through immense networks of information. It can also, occasionally, draw together vast crowds onto the streets, such as we saw in Seattle, Davos, Melbourne and a host of other cities around the world. However, in many ways these assemblages, virtual and actual, are technology driven. ICTs are so convenient, so pervasive and so instrumental, that they almost force these collectivities. To use an overworked cliché, the medium is very much the message in techno-politics. The virtual nature of this form of politics, however, creates a deep contradiction. As ICTs connect and interconnect, they also disconnect. Their instrumentalizing logic disconnects people and the political process from the elements that give it life and power—trust, commitment and responsibility. Disconnected from an organic linkage with people and places and the temporal and social bonds that emerged from it, the momentum of virtual and real-time politics is fatally prone to dissipate as quickly as it begins, a momentary surge of power that cannot sustain itself.

The strong and vibrant "virtual communities" that technophile Howard Rheingold (2000) sees as the new building blocks for civil society become, in this analysis, weak and fatally flawed instruments for social change. They simply do not have the organic strength to effect far-reaching change in society. Virtual politics is also at the same time a sort of simulated politics, a politics based thinly upon a simulacrum that is more *spectacle* than social movement for real change. Croatian philosopher Mario Radovan (2001:110) argues this point well. "Information technology," he writes:

> promotes the image as the dominant form of communication. An image expresses a message almost instantly, and across linguistic, cultural and political

borders far more easily than the printed word does. By promoting the graphic communication, information technology has made an essential contribution to the new *global culture of the present*. Concentrated on the present (and the near future), this new global culture has dislocated itself from "history" understood as a set of stories (often partial) about the past, created by and about specific groups, ideologies and values. Hence we can say that the culture of the information age is essentially *ahistorical*.

Established political parties (in their ostensible role as movements for social change) are not immune from the informationization of politics. The major political parties in all of the Anglo-American economies and beyond have, organizationally and intellectually, synchronized themselves with the temporality of the neoliberal globalization/ICT revolution nexus. In other words, they operate almost on a real-time or short-term basis, a *chronoscopic politics*. The new-style "pragmatic" politician, like the "new breed" university VCs we discussed earlier, considers him or herself in the "business of politics," a business very much like any other, where they need to respond to market signals and be wary of any "negative" market response through their media utterances or policy formulation. For the major parties, especially at election time, "the image" has become the "dominant form of communication." The real-time, short-term focus has major problems for parties of both "left" and "right." Primary among these is that they have disconnected themselves from their histories and their traditional constituencies. As a result they are not sure of what they stand for any more. And so to make themselves "relevant" in this hyper-accelerated political process, they increasingly rely upon focus groups and polling, getting a "real-time fix" on what people are thinking. The trouble with this approach, as Gertrude Stein observed a long time ago, is that "when a man can take a poll and tell what everybody is thinking, that means nobody is really thinking anymore." Not deeply, critically and long-term, anyway.

Tactical Media

Given that the network society is here to stay and is set to insert itself deeper into the political process, both institutional and radical, how do we construct a democratic politics that more truly reflects its digital ecology?

In the early 1990s, only a decade or so ago, the word "Internet" wasn't even in the dictionaries. Similarly, the term "network society" was in limited currency and hadn't yet made the Collins or the Webster's. In

2002 a remarkable document appeared on the Internet that gave a stark indication of how much things have changed. It came from the Office of the President, and was written by "The Presidents' Critical Infrastructure Protection Board" in Washington, DC, and was, therefore, representative of highest-level modes of thinking in the USA. Entitled *The National Strategy to Secure Cyberspace*, it argued that ICTs have changed US society almost completely, and so what is considered "critical" to national security needs to be rethought:

> By 2002, our economy and national security are fully dependent upon information technology and the information infrastructure. A network of networks directly supports the operation of all sectors of our economy—energy (electric power, oil and gas), transportation (rail, air, merchant, marine), finance and banking, information and telecommunications, public health, emergency services, water, chemicals, defense industrial base, food, agriculture, and postal and shipping (p. 9).

You don't get much more unambiguous or comprehensive than that. ICTs and the rapid evolution of the network society have changed everything. Crucial to our discussion, however, is that whilst the world has changed, the *mechanics of politics* have remained very much the same (Beck, 2000). Nineteenth-century institutions and practices such as parliaments, voting systems, geographic representation of constituencies, machine party politics, structured party organizations, branch meetings and so on are still the fundamental mechanisms of organized, institutional political life. Non-institutional politics, the oppositional politics of "the marginalized" and "the oppressed," also follow the traditional formats of street demonstrations, strikes, rioting, petition gathering, and letter writing, etc. In his Internet essay "Globalization, Technopolitics and Revolution," cultural studies theorist Douglas Kellner (2001:5) writes that ICTs "are of essential importance *already* for labour, politics, education and social life, and that people who want to participate in the public and cultural life of the future will need to have [ICT] access and literacy." This is undoubtedly correct, but the problem is that politics, both institutional and oppositional, are trying to incorporate ICTs and the network society into nineteenth-century structures. The result, as I discussed previously, is the "disconnection" of institutional party politics from the mass of the people they claim to represent and the ultimate ineffectuality of mass movements based upon ICTs such as the anti-globalization movement.

The network society is still a work in progress and a site for political contestation, as *The National Strategy to Secure Cyberspace* implies. However, old political structures and practices upgraded to a digital format won't work. The digital environment created in the age of information does not connect easily with politics from the age of industrialism. Politics is still about power, but power in the network society works differently. Scott Lash argues in his *Critique of Information* (2002) that in the network society "power is everywhere" and is "no longer so much something that takes place within elements of the system, between capitalists and proletarians, but instead has to do with exclusion from the system" (p.75). For those excluded, and for those who are trying, as Kellner urges, to effect change from within, using ICTs as weapons, there is no longer a definable locus of effective power which peaceable types can send a petition to or others can throw a crateful of Molotov cocktails at; there is no longer a Winter Palace to storm and thus change the whole balance of power. The anti-globalization mass gatherings to protest the policies of the WTO or whatever in this context are, as I noted previously, more media/political spectacle than effective means for change. Large and unwieldy, and held in tension only through the loose digital ties of the network, *like* the network, they are amorphous, undisciplined and fluxual things that may be dissipated by the constantly shifting actions of the network itself or by mace, water cannon, truncheon and CS gas canister on the streets.

How do we get around this? How is it possible to effect real change in the politically contested site that is cyberspace? How would it be possible and practicable to use ICTs to work within the logic of ICTs and to effect a form of politics more suited to them and the changed social, cultural and political world they have created? Dutch media theorist Geert Lovink has along with others advocated effecting change through use of what they call *tactical media*. Geert Lovink and David Garcia (1996) argue that "Tactical Media are what happens when the cheap 'do it yourself' media...are exploited by groups and individuals who feel aggrieved by or excluded from the wider culture." These groups and individuals must use ICTs "tactically" instead of "strategically" and be "rebellious users" who are able to develop the skills of the tactician or guerrilla in partisan warfare, the "aesthetics" of "poaching, tricking, reading, speaking, strolling, shopping, desiring. Clever tricks, the hunter's cunning, maneuvers, polymorphic situations, joyful discoveries, poetic as well as warlike." Tactical means first and foremost *local*—face-to-face and spatially proximate. Tactical media would comprise local groups and individuals who reject the "classic rituals of the underground and alternative scene" and instead work on the principle of "flexible response," but at the same time building and maintaining elements of trust and commitment.

Tactical media means provisionality, that is, being able to work with different media and in different contexts and at different times in response to the different challenges that arise. Using tactical media does in fact describe what many groups within the anti-globalization movement actually do. I think, however, these are the minority, and certainly many within the movement such as labor unions, minor political parties, socialists, democrats, extreme right-wingers and so on use ICTs in the context of classic political methods and ways of thinking. The methods advocated by Lovink, Garcia and others must become the prevailing logic within politics before politics can be said to have entered the Information Age. This can happen, primarily because the evolving network society is itself inherently "flexible" and "tactical" and not strategic or rigid. In their book *A Thousand Plateaus* (1987), Gilles Deleuze and Felix Guattari use the metaphor of the "rhizome" (a sort of grass, the horizontal underground stems that link and connect above-ground shoots), which acts as an excellent way to think about the reconstitution of politics in the network society. They contrast the rhizome with the tree, which for them represents hierarchy and domination and metaphorically represents nineteenth-century political culture. Rhizomal organizations, so to speak, would be centerless and non-hierarchical, reflecting the more "natural" pattern of geo-organic development and reflecting also the information ecology created by ICTs. Applying this metaphor to humans, Deleuze and Guattari write that: "Many people have trees growing in their heads, but the brain is more like a grass than a tree" (1987:17). And with obvious applicability to the network society, they continue the metaphor by adding that "any point of a rhizome can be connected to any other, and must be. This is very different from the tree or root, which plots a point, fixes an order" (p.7); and moreover, "the rhizome is an acentered, nonhierarchical, nonsignifying system without a General and without an organizing memory or central automaton, defined solely by the circulation of states" (p.21).

Clearly, the metaphor of the rhizome is a useful way to think about how tactical media might work and develop as a political instrument. However, tactical media are something that has to be *learned*, skills to be acquired. The hardware and software don't come with instructions on how to use them as politically subversive or *really* socially creative tools. Governments and industry make much of the idea that people need to be technologically literate—but for instrumental reasons of productivity and profitability. For most people, the rationale underpinning the desire to become technologically literate or expert is to secure a job or a promotion or start a new business. They learn instrumentally because neoliberal capitalism functions instrumentally. But we also need to be technologically literate if we are to forge a new politics in the Information Age. In the spirit of tactical media, though, we need to become techno-*savvy*, to embrace new media and become proficient and "poetic" and "warlike" in their uses. There is another element to this savviness. To be able to use media tactically and as a weapon for the construction of a new politics, and instruments of a new way of thinking, people and organizations need to become *media-savvy* too. So highly mediated is the network society that, for all intents and purposes, everything is media. As Lash (2002:66–67) argues: "what was once 'society' is just as much media as it is society...and what was once 'culture' is just as much media as culture." Technology thus becomes the "linchpin," as Steve Jones (2002b) has termed it, between media and culture—and so to be politically effective, to become media tacticians, we must be able to understand and control the media we use. It is here we must work, I believe, toward the development of a new political culture based upon the "linchpin" of media and culture, toward one that has shorn itself of those ossified elements of nineteenth-century political culture that are no longer relevant. There can be no blueprint for this. We need to identify those outmoded elements and replace them with new ones, and this can only occur through the political struggles that take place in their different contexts across the interlinked, and ultimately interdependent, network society.

The central problem, of course, is that this techno-media savvy, or Internet "cultural competence" as Lovink elsewhere terms it (2002), comes from being able to critique the technology and its media, cultural and political context in our neoliberal society dominated by instrumentalized ICTs. The big battles of the twenty-first century will be over who controls cyberspace and on whose terms. Ordinary people, denied thus far the chance to participate in the network society in anything much more than instrumental terms or as consumers of its entertainments, have a lot of catching-up to do. Consequently a grassroots construction of a *new* politics for the Information Age would be the vital first step.

Chapter Thirteen

Amor Fati (Embrace Your Fate)

How serious is the malady? I've made it sound pretty serious throughout much of this discussion, principally because I think that it is. The "acceleration of just about everything" has gone into overdrive. Increasing connectivity has created an artificial, digital ecology that generates its own real-time temporality. As the information technology revolution spreads deeper and wider, as it surely will, more and more of us will inhabit this constant present, synchronizing our habits and our thinking to its metronomic pulse. Constant or repeated exposure to the information ecology begins to affect us after a while. The symptoms we've discussed. Inside the "buzz of the flickering present" that is the information ecology, we begin to shape our thinking to its temporality and synchronize (or try to) our lives to its demands. Increasingly suspended in real-time, we lose a sense of the past and of the future. The information ecology's instrumental and action-oriented dynamic means that we get used to dealing in real-time with the issues and tasks at hand. Adults coming to the information ecology through work necessity and/or through the general avalanche of connectivity that has characterized the information technology revolution are in danger of "forgetting," as Adorno put it, lost in what he might have also called a "connective dementia" where we forget the modes of thinking, habits of minds, and ways of processing, classifying and digesting information that came with clock time temporality and the world it has metered for over two hundred years. Young people who are weaned on the Internet, Playstation, cell-phones, Bluetooth, ICQ, email, SMS and so on have little or no exposure to other ways of thinking, being and seeing. Real-time for them is *real time*, self-evident, natural, preferable and efficient time. And as we have discussed at some length, the universities are no refuge for contemplation, for the development of a capacity for critical thinking and for the exploration of new ideas that don't have direct commercial applicability—not in any world-changing ways, that is. They perpetuate and exacerbate the problem massively.

"The problem" I just mentioned should be in the plural. Primarily they are technological, cultural, social and political, and I believe that remedies must be based upon these realms. Pre-eminently this must come through a reconstituted *politics* that would help reshape the other realms upon a more democratic and inclusivist foundation. Thomas Hylland Eriksen's book *Tyranny of the Moment: Fast and Slow Time in the Information Age* (2001) ends with a chapter entitled "The Pleasures of Slow Time," which is essentially a personal plea for people to make a determined effort to resist the speeding-up of every aspect of our lives, to "protect slowness." He suggests we can read fewer emails, switch off the cell-phone, throw away the answering machine, make time for the family, read journals instead of newspapers a couple of times a week, go to concerts every couple of weeks, and choose to live in the present whenever it suits. All laudable practices up to a point, but for me this is not good enough. This is politics at the personal level and remains at the personal level. It implies a boycott of information technologies, which implies again that we must somehow avoid or run from them. It is little more than individualistic wishful thinking in the face of powerful systemic pressures that drive people to time-starvation. So what do we do?

There is a quotation, a pithy little aphorism that I come across time and time again on the Internet. It seems to be one of those quotations that simply get replicated over and over—at the beginning or end of an online essay or as a neat little sign-off quote underneath someone's email. It is a great quote and works as an easy device, a cool little one-liner to show where one is coming from. I've never been able to find where its originary point is, the book it comes from, but it's attributed to Frantz Fanon, a psychiatrist and theorist of colonialism who died in 1961. It goes: "A community will evolve only when the people control their means of communication." As I said, it is a wonderful little saying, expressing a deep social insight at the same time as a commonsense obviousness. But how do we do this? How do we make our cyberspace communities *real communities*, as Fanon would have envisaged them? We must begin by looking at the nature of the network society and by identifying the challenges it presents to those looking to effect social, cultural and political change.

Scott Lash in his *Critique of Information* (2002) perceives the ultra-pervasiveness and fundamental importance of ICTs in contemporary life in ways similar to the anonymous bureaucrats who compiled the *National Strategy to Secure Cyberspace*. He maintains that ICTs are so indelibly suffused into who we are and what we do, that nothing exists outside "it" (the digital realm) anymore, nothing that can challenge or change the "information order" to any serious degree:

> The global information order has erased and swallowed up into itself all transcendentals. There is no outside space anymore for critical reflection. And there is just as little time. There is no escaping from the information order, thus the critique of information will have to come from inside the information itself (p.1).

For Lash, the pervasiveness of the information order leaves us with a dilemma—and with a choice to make. Either we accept passively the social, cultural, economic and political structures that the neoliberal information order provides for us to exist within, or we turn to Nietzsche's idea of *amor fati* (embrace your fate) as a way of reasserting individual and collective control over our social means of communication (Lash, 2002:10). In other words, if we have no option but to confront an information order, a network society, then we need to seize control of its structuring agents (namely ICTs) and make them work for us, instead of working for the machine ourselves. To "rage against the machine," as to rage against the technologies themselves, is, therefore, a pointless waste of political, cultural and reflexive energy and doomed to defeat. We need to "rage" against the structure of the network society as it is presently disorganized, using the tools the "machine" itself has provided.

The choice of the verb to "embrace" instead of to "use" becomes, in this context, an important distinction. Embrace means to reach out and grab ICTs before they come looking for you, before you are compelled to "use" them at work, at school or at home. Embracing ICTs in daily life for the purposes of individual and collective empowerment would be to approach them and to exploit them with a positive attitude. This "attitude" would be akin to the eagerness with which a fifteen-year-old approaches Playstation or an artist takes to Photoshop or the eagerness with which a drum'n'bassist would grab the latest digital piece of software or hardware that would allow him or her to create more original sounds. Importantly this eagerness to embrace information technology would combine the two dimensions of technological and media "savvy" that we discussed as the prerequisites of becoming a "media tactician." The instrumentalist dimension of techno-savvy would need to be fused with the reflexivity of media savvy—a knowledge of and expertise in how the technology is culturally and politically situated, and an ability to exploit both the technology itself and the context in which it exists in time and space.

I quoted Raymond Williams previously when he argued that forms of social and political domination (in this case, the domination of life by ICTs) by their very nature remain incomplete and that domination always "involves a limitation...of the activities it covers, so that by definition it cannot exhaust all social experience." Consequently, there will always exist "spaces...where alternative acts and alternative intentions" may germinate and flourish (1979:252). In a world, an "information order" as Lash terms it, that is dominated by ICTs to the extent that there "is no escaping from it," where to be on the "outside" is to be excluded and powerless, where do we find these spaces of difference? Geert Lovink has provided a way to think about the creation of new spaces in his book *Dark Fiber* (2002). Indeed, Lovink argues that we are at a particular point in history: existing in that liminal space, that open-ended interregnum between the end of the dotcom boom and with it the end of a good deal of cyber-utopianism—and the reality of the network society as an entrenched reality. This is the space where, effectively, no one is in charge right now, where power waxes and wanes and flows to and from where the networked money markets anarchically direct it (Beck,1998). This space presents us with many opportunities, but we need to be able to identify and make positive use of these. "Dark Fiber," is both the title of the book and a metaphor. Computer engineers use the term to describe the optical fiber infrastructure that is currently in place but is not being used. The metaphor actually works on two levels. The first is the almost literal level of there being the space, the bandwidth, the ICTs in general available and not being used to anything like their full potential or capacity. Computers are everywhere, and the cables, the links, the repeaters, the satellites, and the telecommunications systems are so dense that plenty of utilizable "dark fiber" exists as "space" throughout the networks of networks for individuals to "embrace" and use as tactical media. Second, this "dark fiber" space can be seen as the technology through which *new* alternative spaces can be "produced," the social "production of [digital] space" in Lefebvre's (1991) sense of the term—within the network society itself.

What Lovink advocates is the emergence of tactical media practices that combine aesthetics, politics and theory. Media theorist McKenzie Wark (2002) writes that Lovink is aware that whether the media tactician comes from theory, art, or activism, what counts is the ability to combine attributes of all three. From the politics comes the art of compromise, of addressing different people directly about things that affect them, and working with people within an autonomy that respects differences without fetishizing them. From the art comes the politics of how languages work, of how to seed awareness of communication, and to do it in appropriate forms. From the theory comes both the art and politics of relating the conjunctures of the moment to history.

However, I argue that politics is at the forefront of this form of change making, a political *strategy* that uses art and the language of aesthetics for political ends (as well as for their own sake) and uses theory to inform the politics and to be informed by it in its turn in the light of practical experience.

Lovink's book draws upon case studies from various projects around the world where ICTs and the Internet, in particular, are being used to create new spaces for communication, spaces for political activism, for discussion, art, criticism and, centrally, as "tools for resistance" (e.g., www.indymedia.org).

It is not difficult to find this sort of stuff all over the Internet with the buzz emanating from possibly hundreds of towns and cities in every continent and nearly every country on the planet. And, of course, these people talk with each other, inhabit each other's spaces and contribute to and take from each other's projects and community building in an ongoing and growing dynamic. Relatively speaking, though, it's a drop in the ocean. In terms of raw numbers, those who are consciously involved in this sort of thing in a serious and sustained way would number in the tens of thousands. To be effective, self-conscious activism of this sort must generalize into the practice of everyday life if the logic of an instrumentalized and accelerated network society is to be changed. However, considering the network society is barely a decade old, we have already taken the first and most important steps. More and more people realize what it is we are dealing with and what needs to be done. The rewards, potentially, are great. In a short time millions of us have adapted to the digital, chronoscopic treadmill; millions more, those from a younger generation, especially, have known little else. The scale of the desired ontological reorientation away from seeing ICTs as an instrumental tool is, therefore, vast, and the self-organized release from digital drudgery would truly constitute a revolution.

Turning around our lives need not mean having to accomplish the titanic task of doing away with capitalism (however desirable this may be). It would mean a *reforming* of capitalism through the revolutionizing of its means of production, a task that is much more conceivable and achievable in this era where the ruling ideology of capitalism has no serious challenges. It would mean, as I said, the theory and practice of the self-conscious media tactician spilling over into everyday life. It would mean the use of ICTs, not as an alternative to face-to-face and localized human interactions, but as a way of creating and strengthening this most human of practices and, in doing so, maintaining and building the dimensions of trust and commitment that are needed to take the larger political project forward. This project would mean using tactical media to produce new *spaces* and *temporalities* that are explicitly concerned with working against the unsustainable "acceleration of just about everything" that our present neoliberal configuration of the network society has generated, showing that alternatives are possible and workable—in one's job, study, home life and family life, showing that digital temporality need not mean the unerring and unbending meter of real-time but that an infinite number of temporalities can exist within the network society to correspond with a diversity of local and contextual cultures, societies and polities.

The first *major* task of reformism, one from which much else would be possible, is the breaking of the nexus between information technologies and the neoliberal project. And this seems less and less of an impossible dream, so fraught and fragile has the entire neoliberal project been. Indeed, the beginning of the twenty-first century witnessed the first signs of a major hangover, resulting in the excesses of the rise and domination of neoliberalism over the last quarter of the twentieth century. The simple truth that free markets and deregulation more than likely mean that the powerful will trample the weak and cheat and steal their way to domination began to emerge in spectacular style with the collapse of companies like Enron and WorldCom. CEOs were filmed in handcuffs being bundled into police cars. Tens of thousand of jobs were lost; billions of dollars in pensions, wages and entitlements evaporated, and the neoliberal fetish of the market was responsible for much of this (Frank, 2002). Neoliberalism is rotten at its core, and notwithstanding our diminishing concentration span in a fast-moving, digitized life, people are beginning to realize this. The hundreds of thousands across the world who have lost their New Economy jobs or their life savings or have seen their share portfolio shrink to worthlessness are beginning to realize this; the millions across the world who read about the colossal scope of the frauds that are being perpetrated by big business think about their own jobs and pensions and stockholdings and also begin to realize that possibly root-and-branch market deregulation may not the nirvana it was portrayed as in the 1980s and 1990s.

In reality, the nexus between neoliberalism and the ICT revolution is a weak one, an ideological hold over a form of technological development that seems to be unraveling. As it unravels more, it is vital that people make the right connections about how and why free markets cannot be the solution and, more importantly, about how ICTs can contribute much more to our lives and our communities than their restricted and instrumental application today. Given the right conditions, universities would have a role to play here. They could become in actuality once more, as Delanty sees them now, "key institutions" where "knowledge, culture and society interconnect" (2001:vii). The vacuum that is likely to emerge from an unraveling of neoliberal ideology could be filled with a diversity of articulations regarding why the rule of the market has been such a disaster and pointers to possible ways forward.

Most importantly of all, the broken nexus between neoliberal globalization and information technologies would shatter the pervasive information ecology and the real-time constant present that it generates and sustains. People would then be able to regain some control over time once again, using it instead of being driven by it. The network society, a society which is here to stay whether we like it or not, could consist of various temporalities and spaces that we can inhabit and move through in different aspects of our lives, our work, our leisure pursuits, our family and our communities. The reclamation of half-forgotten temporalities that we only intuit today could open up new and fulfilling dimensions to culture, to innovation, to creativity and to technology. Clock time, real-time, family time, reading time, working time, thinking time and doing nothing time can have the chance to once more "interpenetrate and permeate our lives" as Barbara Adam put it, contributing to a holistic existence that evolves around a balance. In such a culture, economy and society, my son Theo will be able to immerse himself in a Playstation game for two or more hours and then lose himself in a book. These temporalities would not have to cancel each other out. He would be able to work creatively in the pressured environment of real-time that is not going to disappear, or, if he chooses, do his creative work with a canvas and paint, or a video camera, or whatever. At university, if he chooses to go, he will have the time and the freedom to study philosophy or advertising—or both together—the point being that he would be able to do them both properly and to the best of his ability. He could send an email to a friend on the other side of the world to say that he's just sent to her by "snail mail" a charcoal rendering of the JPEG digital self-portrait she sent to him over the Internet the day before. Theo's projects could be completed, friendships sustained, contacts maintained and ideas thought through to their natural conclusions in ways that are either rapid and off-the-cuff or slow and considered. He would have the option do both, or none, and then move on in his own time. In short, he could have the times of his life.

Bibliography

Adam, Barbara. *Timewatch: The Social Analysis of Time*. Cambridge: Polity, 1995.
——. *Timescapes of Modernity: The Environment and Invisible Hazards*. London: Routledge, 1998.
Adorno, T. W., and Horkheimer, M. *Dialectic of Enlightenment*. London: Verso, 1944/1986.
Apple, Michael. "Review of 'Education and the Economy in a Changing Society' by the OECD 1989." *Comparative Education Review* 36 (1), 1994.
Armstrong, Philip et al. *Capitalism Since World War II*. London: Fontana, 1984.
Astin, Alexander. "The Changing American College Student: Thirty-Year Trends, 1966–1996." *Review of Higher Education* 21 (2) 1998.
Barber, Benjamin. "The New Communications Technology: Endless Frontier or the End of Democracy." *Constellations* 4 (2) 1997.
Barnett, Ronald . *Higher Education. A Critical Business*. Buckingham: Open University Press, 1997.
Baum, Ruth. "The Educated Investor." *Kansas City Star Online*. www.kcstar.com 15 July 2001.
Beck, Ulrich. "The Politics of Risk Society" in Jane Franklin, ed. *The Politics of Risk Society*. Cambridge: Polity, 1998.
——. *What Is Globalisation?* Oxford: Blackwell, 2000.
Bell, Daniel. *The End of Ideology: On the Exhaustion of Political Ideas in the Fifties*. New York: Free Press, 1962.
Berman, Marshall. *All That Is Solid Melts into Air: The Experience of Modernity*. London: Verso, 1982.
Boulter, David J. *Turing's Man: Western Culture in the Computer Age*. London: Duckworth, 1984.
Bourdieu, Pierre. "Social Space and Symbolic Power" in *In Other Words: Essays Toward a Reflexive Sociology*. Cambridge: Polity, 1990.
Carlson, S. "Computer-Savvy Students Perform Poorly on Handwritten Compositions, Study Finds." *Chronicle of Higher Education* 18th Aug. 2000.
Castells, Manuel. *The Rise of the Network Society*. Oxford: Blackwell, 1999.
——. *The Internet Galaxy*. New York: Oxford University Press, 2001.
Computer Industry Almanac Inc. www.c-i-a.com/pr032102.htm 2002.
Currie, Jan. "Introduction" in Jan Currie and Janice Newson, eds. *Universities and Globalization: Critical Perspectives*. London: Sage, 1998.
Davidow, William, and Malone, Michael. *The Virtual Corporation: Structuring and Revitalizing the Corporation for the 21st Century*. New York: HarperCollins, 1992.
Davies, Norman. *Europe: A History*. London: Pimlico, 1997.
Debord, Guy. *The Society of the Spectacle*. Detroit: Black and Red, 1983.
Delanty, Gerard. *Challenging Knowledge: The University in the Knowledge Society*. Buckingham: Open University Press, 2001.
Deleuze, Gilles, and Guattari, Felix. *A Thousand Plateaus: Capitalism and Schizophrenia*. Minneapolis: University of Minnesota Press, 1987.

Diani, Mario. *Social Movement Networks: Virtual and Real, Information, Communication and Society* 3 (3) 2000.
Dorsey, James. "Going Mobile." *Wall Street Journal Online*. www.wsj.com/public/current/articles/SB984072265634007065.htm 2001.
Eagleton, Terry. *Ideology*. London: Verso, 1991.
EMC. www.emc-database.com/website.nsf/index/pr020319#this-page 2002
Eriksen, T. H. *Tyranny of the Moment: Fast and Slow Time in the Information Age*. London: Pluto, 2001.
European Commission. *The Globalisation of Education and Training*. http://europa.eu.int/comm/education/global.pdf 2000.
Fabian, Johannes. *Time and the Other: How Anthropology Makes Its Object*. New York: Columbia University Press, 1983.
Ferguson, Wallace. *Europe in Transition*. Boston: Houghton Mifflin, 1962.
Florida, Richard. "The Role of the University: Leveraging Talent, not Technology." *Issues in Science and Technology Online*. www.nap.edu/issues/15.4/florida.htm 1999.
Forrester Research: http://ecitizen.mit.edu/classes/presentations/presentations/B2B/sld003.htm 2001.
Foucault, Michel. *The Archaeology of Knowledge*. New York: Pantheon, 1972.
Frank, Thomas. *One Market Under God: Extreme Capitalism, Market Populism and the End of Economic Democracy*. London: Secker and Warburg, 2001.
———. "The Enron Outrage." *Salon.com*. http://www.truthout.org/docs_01/12.16E.Enron.Outrage.htm, 2002
Gare, Arran. *Nihilism Inc.: Environmental Destruction and the Metaphysics of Sustainability*. Sydney: Eco-Logical Press, 1996.
Gates, Bill. *The Way Ahead*. London: Viking, 1995.
GCC Yearbook. *Global Civil Society Yearbook, 2001*. Centre for Civil Society ed. Centre for the Study of Global Governance, Oxford: Oxford University Press, 2001.
Gehring, John. "Student Jobs Hurt Math, Science Scores." *Education Week*. www.edweek.org/ew/ewstory.cfm?slug=11work.h20 2000.
Genuine Progress Index (GPI Canada). www.gpiatlantic.org/gpinewsletter.shtml 2000.
Gibbons, M., Limoges, C., Nowotny, H. et al. *The New Production of Knowledge*. London: Sage, 1994.
Gibson, Dennis, and Hatherall, William. "Reflections on Stability and Change in Australian Higher Education" in John Sharpman and Grant Harman, eds. *Australia's Future Universities*, Armidale, NSW: University of New England Press, 1997.
Giddens, Anthony. "Institutional Reflexivity and Modernity" in *The Polity Reader in Social Theory*. Cambridge: Polity, 1994.
———. *Runaway World: How Globalisation Is Reshaping Our Lives*. London: Profile, 1999.
Gilbert, Alan. "Universitas 21 Offers Virtual Edge." http://www.unimelb.edu.au/ExtRels/Media/UN/archive/multimediatwo/universitasoffersvirtualed.html 2000.
Gleick, James. *Faster: The Acceleration of Just About Everything*. New York: Abacus, 1999.
Haahr, Mads. "Dreams of an Accelerated Culture." *Crossings* 1 (1) http://crossings.tcd.ie/issues/1.1/Haahr/ 2001.
Habermas, Jürgen. *The Theory of Communicative Action*, vol. 2. Boston: Beacon, 1987.

Hamza, Mohammad, and Alhalabi, Bassem. "Technology and Education: Between Chaos and Order." *First Monday* 4 http://www.firstmonday.dk/issues/issue4_3/hamza/ 1999.
Harvey, David. *The Limits to Capital*. Oxford: Blackwell, 1983.
———. *The Condition of Postmodernity*. Oxford: Blackwell, 1989.
———. *Justice, Nature and the Geography of Distance*. Oxford: Blackwell, 1996.
Hassan, Robert. "The Space Economy of Convergence." *Convergence* 6 (4) 2000.
Hewitt, Paul. *Conceptual Physics: A New Introduction to Your Environment*. Boston: Little, Brown, 1974.
Hoad, Jeremy. "Servants of the Economy." *The Guardian Online* http://education.guardian.co.uk/higher/story/0,5500,370008,00.html 2000.
Hörning, Karl H., Ahrens, Daniela, and Gerhard, Annette. "Do Technologies Have Time?" *Time and Society* 8 (2) 1999.
Horovitz, Bruce. "College Life a Whole New E-world," *USA Today*, August 19th 1999.
Hughes, Robert. *The Fatal Shore*. New York: Knopf, 1986.
Jacoby, Russell. *The End of Utopia*. New York: Basic, 1999.
Jameson, Fredric. *Postmodernism, or, the Cultural Logic of Late Capitalism*. London: Verso, 1995.
Jones, Steve. "The Internet Goes to College." Pew Internet and the American Life Project http://www.pewinternet.org/reports/toc.asp?Report=71 2002a.
———. Personal communication, 24 Sept. 2002b.
Jonscher, Charles. *The Evolution of Wired Life*. New York: John Wiley and Sons, 1999.
Kellner, Douglas. "Globalization, Technopolitics and Revolution." www.gseis.ucla.edu/faculty/kellner/kellner/html 2001.
Kern, Stephen. *The Culture of Time and Space 1880–1914*. London: Weidenfeld and Nicolson, 1983.
Klein, Naomi. "The Vision Thing." Coalition for Global Solidarity and Social Development. http://www.protest.net/view.cgi?view=1785 2000.
Kolko, Joyce. *Restructuring the World Economy*. New York: Pantheon, 1988.
Langtry, Bruce. "Ends and Means in University Policy Decisions" in Tony Coady, ed. *Why Universities Matter*. Sydney: Allen and Unwin, 2000.
Lasch, Christopher. *The Revolt of the Elites and the Betrayal of Democracy*. New York: W. W. Norton, 1995.
Lash, Scott. *Critique of Information*. London: Sage, 2002.
Lash, Scott, and Urry, John. *The End of Organized Capitalism*. Cambridge: Polity, 1987.
Lefebvre, Henri. *The Production of Space*. Oxford: Blackwell, 1991.
Lehmann, Nicholas. "A Fool's Goal." *Forbes.com*: http://www.forbes.com/asap/1998/1130/281.html 1998.
Levy, Steven. "Time to Act." *Newsweek* 9 July 2001.
Lingard, Robert, and Rizvi, Fazal. "The OECD and Australian Education" in Jan Currie and Janice Newson, eds. *Universities and Globalization: Critical Perspectives*. London: Sage, 1998.
Lobkowicz, Nikolaus. "Man, the Pursuit of Truth and the University" in John W. Chapman, ed. *The Western University on Trial*. Berkeley: University of California Press, 1983.
Lovink, Geert. *Dark Fiber*. Cambridge, MA: MIT Press, 2002.

Lovink, Geert, and Garcia, David. http://www.waag.org/tmn/abc.html 1996.
Lowen, Rebecca S. *Creating the Cold War University*. Berkeley: University of California Press, 1997.
Lyotard, Jean-François. *The Postmodern Condition: A Report on Knowledge*. Manchester: Manchester University Press, 1979.
MacDonald, Robert. and Coffield, Frank. *Risky Business? Youth and the Enterprise Culture*. London: Falmer, 1991.
MacIntyre, Alasdair. *After Virtue: A Study in Moral Theory*. Notre Dame, Ind.: Notre Dame University Press, 1984.
Marcuse, Herbert. *One Dimensional Man*. London: Sphere, 1968.
Marginson, Simon, and Considine, Mark. *The Enterprise University: Power, Governance and Reinvention in Australia*. Cambridge: Cambridge University Press, 2000.
Martin, Elaine. *Changing Academic Work: Developing the Learning University*. Buckingham: Open University Press, 1999.
Marx, Karl. *Capital*. Harmondsworth: Penguin, 1867/1967.
Marx-Engels. *Selected Works*. Moscow: Progress, 1975.
McInnis, Craig. "Signs of Disengagement: The Changing Undergraduate Experience in Australian Universities." Inaugural Professorial Lecture. http://www.cshe.unimelb.edu.au/downloads/InauguralLecture.pdf 2001.
McLuhan, Marshall. *The Medium Is The Massage: An Inventory of Effects*. Corte Madera: Gingko, 1967/2001.
Milne, Seamus. *The Enemy Within*. London: Verso, 1994.
Moisan, Dorothee. "Web Users Fight Advertising Emails." *Nando Times* www.nandotimes.com/technology/story/15796p-292843c.html 2001.
Morris, Meaghan and McCalman, Iain. "Knowing Ourselves and Others." *Public Culture* 3 (1) www.asap.unimelb.edu.au/aah/research/review/cl_morris.html 2000.
Mulhauser, Dana 2001. "New Virtual University in Virginia Would Let Students Create Their Own Curriculum." *Chronicle of Higher Education* 24 July 2001.
Mumford, Lewis. *Technics and Civilization*. London: Routledge and Kagan Paul. 1934/1967.
Negroponte, Nicholas. *Being Digital*. New York: Vintage, 1995.
Nielsen Net Ratings. http://209.249.142.22/press_releases.asp?country=north+America 2001.
Noble, David F. "Digital Degree Mills: The Automation of Higher Education." *First Monday* 3 http://www.firstmonday.dk/issues/issue3_1/noble/ 1998.
Nowotny, Helga. *Time: The Modern and Postmodern Experience*. Cambridge: Polity, 1994.
Ong, Walter. *The Presence of the Word: Some Prolegomena for Cultural and Religious History*. Minneapolis: University of Minnesota Press, 1967.
Oppenheimer, Todd. *TheAtlanticOnline* www.theatlantic.com/isues/97/computer.htm 1997.
Papadopolous, George. *OECD 1960–1990: The OECD Perspective*. Paris: OECD, 1994.
Peppers, Don, and Rogers, Martha. *The One to One Future*. New York: Currency Doubleday, 1993.
Perera, Rick. "Mobile Phone Sales up by 46 Percent Last Year." IDG News Service http://www.security-informer.com/english/crd_market_427647.html 2001.

President's Information Technology Advisory Committee PITAC. "Ten Critical National Challenge Transformations." www.ccic.gov/ac/report/ 2000.

Polster, Claire, and Newson, Janice. "Don't Count Your Blessings" in Jan Currie and Janice Newson, eds. *Universities and Globalization: Critical Perspectives.* London: Sage, 1998.

Postman, Neil. *Technopoly: The Surrender of Culture to Technology.* New York: Vintage, 1993.

Press, Eyal. and Washburn, Jennifer. "The Kept University." *TheAtlanticOnline* http://www.theatlantic.com/issues/2000/03/press.htm 2000.

Purser, Ron. "The Coming Crisis in Real Time Environments: A Dromological Analysis." online.sfsu.edu/~rpurser/revised/pages/DROMOLOGY.htm 2000.

Radovan, Mario. "Information Technology and the Character of Contemporary Life." *Information Technology, Communication and Society* 4 (2) 2001.

Reich, Robert. *The Work of Nations.* New York: Random House, 1991.

Rheingold, Howard. *The Virtual Community: Homesteading on the Electronic Frontier.* Cambridge, MA: MIT Press, 2000.

Rifkin, Jeremy. *The Age of Access.* London: Penguin, 2000.

Rochlin, Gene. *Trapped in the Net: The Unanticipated Consequence of Computerization.* Princeton, NJ: Princeton University Press, 1997.

Rock, Arien, and Mack, Irvine. *Inattentional Blindness.* Cambridge, MA: MIT Press, 1998.

Ross, Andrew. *Strange Weather: Culture, Science and Technology in the Age of Limits.* London: Verso, 1991.

Rossetto, Louis, "Change is Good" *Wired Magazine*, Vol.6 01, 1998

Roszak, Theodore. *The Cult of Information*, Berkeley: University of California Press, 1986.

Rüegg, Walter. "A Conversation About the Humanities" in John W. Chapman, ed. *The Western University on Trial.* Berkeley: University of California Press, 1983.

Schevitz, Tanya. "Point, Click, Plagiarize: Web Site Nabs UC Berkeley Students Stealing from Net." *San Francisco Chronicle*, 5'November 1999.

Schiller, Dan. *Digital Capitalism: Networking the Global Market System.* Cambridge, MA: MIT Press, 1999.

Showalter, Elaine. *The Female Malady: Women, Madness, and English Culture, 1830–1980.* New York: Pantheon, 1985.

Slaughter, Sheila, and Leslie, Larry, L. *Academic Capitalism.* Baltimore: Johns Hopkins University Press, 1997.

Smith, Adam. *An Inquiry into the Nature and Causes of the Wealth of Nations.* New York: Modern Library, 1776/1965.

Stewart, F. "Meeting the Online Challenge." *Australian Financial Review*, 21 March 2001.

Swain, Hilary. "A Job's Just Part of the Course." *Times Higher Education Supplement*, 30 Oct. 1998.

Tapscott, Don, Ticoll, David, and Lowy, Alex. *Digital Capital: Harnessing the Power of Business Webs.* Boston: Harvard Business School Press, 2000.

Taylor, Frederick W. *The Principles of Scientific Management.* New York: Harper and Brothers, 1911.

The National Strategy to Secure Cyberspace. www.whitehouse.gov/pcipb/cyberstrategy-draft.pdf 2002.

Thompson, E. P. "Time, Work-Discipline, and Industrial Capitalism." *Past and Present* 38, 1967.

Thornton, Emily. "In Asia, Pursuing Western MBAs Without Leaving Home" *Businessweek Online* http://www.businessweek.com/1998/42/b3600015.htm 1998.

Thrift, Nigel. *Spatial Formations*. London: Sage, 1996.

Touraine, Alain. *Critique of Modernity*. Oxford: Blackwell, 1995.

Universitas 21. http://www.universitas.edu.au/introduction.html 2001.

Virilio, Paul. *The Information Bomb*. London: Verso, 2000.

Wark, McKenzie. "Lovink's Dark Fiber." www.rhizome.org 2002.

Weiner, Norbert. *The Human Use of Human Beings: Cybernetics and Society*. London: Sphere, 1968.

Weiser, Mark, and Seely Brown, John. "The Coming Age of Calm Technology" in Peter J. Denning and Robert M. Metcalfe, eds. *Beyond Calculation: The Next Fifty Years of Computing*. New York: Copernicus, 1997.

Whitrow, G. J. *What Is Time?* London: Thames and Hudson, 1972.

Williams, Raymond. *Politics and Letters*. London: NLB, 1979.

Winston, Brian. *Media, Technology and Society*. London: Routledge, 1998.

Index

A

abbreviated thinking, 6, 133–142, 158
academic capitalism, 7, 74, 77
Adam, Barbara, 11,16, 19–20, 175
Adorno, Theodor, 128–129, 169
Alhalabi, Bassem, 88
anti–globalization movement, 159, 164–166
Apple, Michael, 93
ARPANet, 48–49

B

Babbage, Charles, 48,
Baran, Paul, 156
Barber, Benjamin, 155
Barnett, Ronald, 119, 121, 124, 126–27
Beard, George Miller, 38
Beck, Ulrich, 164, 172
Bell, Alexander Graham, 48
Bell, Daniel, 48, 149–150
Berman, Marshall, 151
Blair, Tony, 143
Boulter, David, 13
Bourdieu, Pierre, 153
brand building (in universities), 75, 79, 85–87, 95
Bush, George W., 143
Bush, George, 152
Bush, Vannevar, 3, 48

C

Castells, Manuel, 4, 48, 50, 53
capitalism,
    as political ideology, 48, 77, 92, 148–153, 173–174
    crises of, 14, 39, 44–49, 54, 73, 92, 152–153
    culture of, 54–55, 73, 118, 137–138, 159
    disorganized 155–167
    flexibilization of, 14, 46–47, 50, 55, 75, 77, 116, 118, 156
    golden age of, 44, 46, 48–49, 148, 150
    globalization and, 37–56, 61–63, 73, 77, 91, 93, 137, 158
    restructuring of, 45–49, 54, 74, 81, 91, 115
chronologic politics, 7, 145, 150–153
chronologic time, 5–6, 16, 35, 38–42, 109, 151
chronoscopic politics, 155, 163–167
chronoscopic time, 5–6, 8, 35, 40–42, 55, 64, 104, 123, 145, 153
civil society, 7–8, 35, 89, 92, 104, 153, 159, 162
clock time, 5–7, 11–17, 19–25, 27–28, 30–32, 34–35, 38, 40, 52, 55, 109, 147, 156, 169, 175
    social construction of, 6, 15–20
Coffield, Frank, 144
commodification,
    of culture and society, 73–74, 129, 137
    of time, 27–36
consciousness–changing,
    role of ICTs in, 64–65
Considine, Mark, 7, 74–75, 77, 94
critical thinking, 5, 41, 89, 100–101, 114, 122, 124–129, 135–137, 146, 149–150, 153, 169, 171
cultural competence, 167
Currie, Jan, 73

D

dark fiber, 172
Dawkins Revolution, 94
Debord, Guy, 129, 149–150

Delanty, Gerard, 7, 77, 116–118, 175
Deleuze, Gilles, 166
Diani, Mario, 162

E

Eagleton, Terry, 134
e-commerce, 61
Engels, Friedrich, 147,
Enlightenment, 8, 18, 20, 28, 31, 70–71, 115–116, 145, 150
Enron, 76, 174
Enterprise University, 7, 74, 96, 99, 103–106, 123, 129, 136
ephemeris time, 16
Ericsson, 84
Eriksen, Thomas Hylland, 2, 5–6, 8, 170

F

Fanon, Frantz, 170
Ferguson, Wallace, 70
Florida, Richard, 92–93
Florida, University of, 87
Ford, Henry, 34, 41
Fordism, 14, 34, 39, 44–45, 49, 52, 54, 72, 92, 96–97, 99–100, 117, 150, 152–153, 156–157
Ford Motor Corporation, 34, 84
Foucault, Michel, 73, 115, 146
Frank, Thomas, 63, 87, 141, 174
Franklin, Benjamin, 32
Friedman, Milton, 46
Fukuyama, Francis, 110, 112, 161

G

Galileo, 18
Garcia, David, 165–166
Gare, Arran, 11, 17, 19
Gates, Bill, 2–3, 56, 61
Gibbons, Michael, 115–118
Giddens, Anthony, 59, 152

Gilbert, Alan, 87
Gleick, James, 1–2, 5–6, 33, 37, 109, 159

globalization, 37–56, 62–63, 73, 77, 84, 90–91, 93, 96, 109–110, 113, 118, 123, 125, 129, 136–141, 145, 158–159, 163, 175
Guattari, Felix, 166
Gullers, Peter, 135

H

Haahr, Mads, 139–140
Habermas, Jürgen, 27
Hamza, Mohammad, 88
Hardy, Barbara, 135
Harvard Business School, 4, 78
Harvey, David, 14, 24, 34, 43, 149, 156
Hayek, Friedrich Von, 46
Hewitt, Paul, 16
Hilferding, Rudolf, 155
Horkheimer, Max, 128
Hörning, Karl H., 13
Horovitz, Bruce, 104
Hughes, Robert, 120–121

I

ICTs, 2–3, 5–8, 14, 25, 40–42, 45–57, 63–64, 73, 77, 81, 83–84, 87–88, 90, 105–106, 113, 118, 123, 128–129, 133–134, 137, 139–140, 159–175
ideology, 48, 62, 73, 77, 92, 148–149, 174–175
inattentional blindness, 138
industrial revolution, 3, 6, 13–15, 18, 25, 28, 31–32, 34–35, 38–40, 43–44, 51, 71, 143–145, 150
information ecology, 5–6, 8, 40–42, 55, 59, 61, 81, 105, 109, 113, 118, 125, 134–136, 138, 143, 156–159, 166, 169, 175

instrumental rationality, 127–129, 149, 151, 159
interconnectivity, 2, 4, 40–41, 50–51, 55, 59, 61, 65, 83, 104, 109, 125, 158
Internet, 2–5, 40, 42, 48, 53, 55, 58–64, 79, 83, 105–106, 111, 126, 158–159, 163–164, 167–170, 173, 176

J

Jameson, Fredric, 137
Jones, Steve, 106, 167
Jonscher, Charles, 119, 121
Joseph, Keith, 45

K

Kellner, Douglas, 164–165
Kern, Steven, 52
Klein, Naomi, 159
knowledge production, 13, 41, 77, 80, 92, 96, 109–110, 114–129, 135, 146, 153, 169
Kocka, Jürgen, 155

L

Langtry, Bruce, 98
Lasch, Christopher, 101
Lash, Scott, 155–156, 165–166, 170–172
Leeson, Nick, 141
Lefebvre, Henri, 172
Lehmann, Nicholas, 140
Leslie, Larry, 74, 77
liberal education, 7, 70, 89, 98–101, 107, 122, 125,
Lingard, Robert, 93
Lobkowicz, Nikolaus, 70–71

Lovink, Geert, 165–167, 172–173
Lucent Technologies, 84
Lyotard, Jean-François, 115, 118, 122–123, 136

M

MacIntyre, Alasdair, 134–135
Mack, Irvine, 138
Marcuse, Herbert, 34, 125, 128–129, 148–150, 153, 156
Marginson, Simon, 7, 74–75, 77, 94
Martin, Elaine, 84–85
Marx, Karl, 25, 29–30, 38–39, 147–148, 153
Massachusetts Institute of Technology, 86
McCalman, Iain, 101
McDonald, Robert, 144
McInnis, Craig, 102–103, 105, 113
McLuhan, Marshall, 1–2, 20, 35
media savvy, 167, 171
Microsoft, 85, 103, 140
MIT Media Lab, 55
modernity, 1, 7–8, 39, 41, 52, 116–117, 128, 144–152, 157
Morris, Meaghan, 101
MP3 file sharing, 4
Mulhauser, Dana, 103
Mumford, Lewis, 13
Murdoch, Rupert, 3

N

Naisbitt, John, 80
narratives, 134–136, 151, 157
Negroponte, Nicholas, 55
neoliberalism, 6–7, 29, 45–47, 49, 54, 62, 73–74, 77, 80, 82, 89–90, 92–96, 118, 125, 129, 137, 140–141, 145, 158–159, 163, 167, 171, 174–175

network society, 2, 7, 48, 50, 53, 90, 115, 118–119, 121, 163–167, 170–175
New Economy, 55, 61–62, 85, 87, 91–92, 140, 174
New Right, 45–46, 49, 54
Newman, Janice, 76
Newton, Isaac, 17–19, 116
Nowotny, Helga, 15, 20

O

Ong, Walter, 25, 134
online degrees, 84, 88
Oppenheimer, Todd, 82

P

Papadopolous, George, 93
Peppers, Don, 57, 63, 78
plagiarism, 105–106

Plato's Academy, 69–71
Polster, Claire, 76
Postman, Neil, 14, 140
postmodernity, 117, 122, 136–137, 158
Purser, Ron, 41

R

Radovan, Mario, 162–163
Reagan, Ronald, 45–46
Real-time, 1–3, 5–6, 8, 15, 40–41, 55–56, 84, 101–102, 104, 109, 123, 125, 133–138, 140, 159, 161–163, 169, 174–175
récupération, 129, 150, 152
reflexivity, 5, 42, 72, 82, 88–89, 101, 117–119, 122, 125, 171
Reich, Robert, 92
Rheingold, Howard, 162
rhizome, 166,
Ricardo, David, 29–30
Rifkin, Jeremy, 62, 73, 85, 137

Rizvi, Fazal, 93
Rochlin, 64, 135
Rock, Arien, 138
Rogers, Martha, 58, 63, 78
Ross, Andrew, 133–134
Roszak, Theodore, 54, 80–83
Rüegg, Walter, 71
Russell, Michael, 82

S

Schiller, Dan, 7, 79–80, 90
Seely Brown, John, 40
sensorium, Ong's concept of, 25, 134
Showalter, Elaine, 38
Slaughter, Sheila, 74, 77
Smith, Adam, 17–19, 28–31, 33, 45–46, 71, 85
space, 25, 35, 39–40, 43, 51–52, 55, 117, 118–119, 125, 129, 133–134, 136, 157, 160–161, 171–175
spatial fix, Harvey's concept of, 43
Steele, Ted, 106
Stein, Gertrude, 163
Swain, Hilary, 104
Sweezy, Paul, 156

T

tactical media, 163–167, 172, 174
Tapscott, Don, 61
task oriented time, 30–31
Taylor, Frederick W., 33
Taylorism, 33–34, 98, 156
techno savvy, 167, 171
technological convergence, 47–50, 53–54
technological determinism, 134, 140
techno–politics, 8, 158–166
temporal acceleration, 1–2, 5–6, 14–15, 37–38, 64–65, 81, 89, 101, 104, 109, 115, 118, 139–140, 159, 169, 173

temporal rhythms, 3, 21–25, 27, 28, 30, 33, 51–52, 55, 99–100, 102, 115
temporality, diversity of, 16–17, 20, 22–25
Thatcher, Margaret, 29, 45–46
Thompson, E.P., 27, 30, 32
Thrift, Nigel, 20
timepieces, historical role of, 30
Toffler, Alvin, 80,
Touraine, Alain, 145–146
trust and commitment,
    in politics, 161–167, 174

U

ubiquitous computing, 40, 55
university,
    history of, 69–73
    commercialisation of, 73–79
    informationization of, 79–90
    internationalization of, 83–87
Urry, John, 155–156

V

Virilio, Paul, 5, 15
virtual corporation, 85

W

Wark, McKenzie, 172–173
Weber, Max, 127–128
Weiser, Mark, 40
Wharton Business School, 84
Williams, Raymond, 129, 172
Windows software, 63–64
Winston, Brian, 49
work time, 12–14
WorldCom, 63, 76, 174
Wireless access protocol (WAP), 50, 60, 84

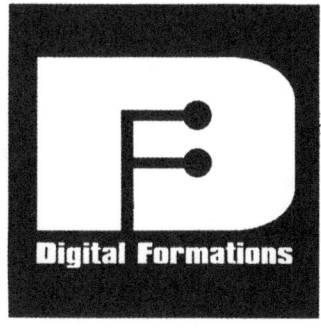

General Editor: **Steve Jones**

Digital Formations is the new source for critical, well-written books about digital technologies and modern life. Books in this series will break new ground by emphasizing multiple methodological and theoretical approaches to deeply probe the formation and reformation of lived experience as it is refracted through digital interaction. Each volume in *Digital Formations* will push forward our understanding of the intersections—and corresponding implications—between the digital technologies and everyday life. This series will examine broad issues in realms such as digital culture, electronic commerce, law, politics and governance, gender, the Internet, race, art, health and medicine, and education. The series will emphasize critical studies in the context of emergent and existing digital technologies.

For additional information about this series or for the submission of manuscripts, please contact:

> Acquisitions Department
> Peter Lang Publishing
> 275 Seventh Avenue 28th Floor
> New York, NY 10001

To order other books in this series, please contact our Customer Service Department:

> (800) 770-LANG (within the U.S.)
> (212) 647-7706 (outside the U.S.)
> (212) 647-7707 FAX

or browse online by series:

> WWW.PETERLANGUSA.COM